PSYCHOANALYTIC INQUIRY

A Topical Journal for Mental Health Professionals

| Volume 12 | 1992 | Number 3 |

Neuroscience and Clinical Science:
Toward an Integration

PSYCHOANALYTIC INQUIRY

Editor-in-Chief

JOSEPH D. LICHTENBERG, M.D.

Editors

MELVIN BORNSTEIN, M.D.   **DONALD SILVER, M.D.**

First published 1992 by Routledge
605 Third Avenue, New York, NY 10017
2 Park Square, Milton Park, Abingdon, Oxon OX14 4RN

Routledge is an imprint of the Taylor & Francis Group, an informa business

ISBN 13: 978-0-88163-946-9 (hbk)
ISSN 0735-1690

Prologue

SOME 14 YEARS AGO NOW, a preeminent neuroscientist with a longstanding interest in psychoanalysis and derivative psychotherapies reopened in rather dramatic fashion a line of inquiry that Freud early on (1895) had envisioned, explored, but then abandoned as premature. In a 1978 lecture to a Harvard audience, Eric R. Kandel (1979) suggested that because of advances in basic research the time had perhaps arrived to begin again to sketch connections between data relating to information-dependent neural plasticity and the cellular neurobiology of learning, on the one hand, and, on the other, clinical findings about the origin of and the psychological treatment of neurosis and character disorder.

Since the time of Kandel's first formal steps toward interdisciplinary integration, the psychiatric and psychoanalytic literatures have expanded to include many additional attempts to meld insights from laboratory and consulting room. The five essays in this issue of *Psychoanalytic Inquiry* both review the progress to date and, more relevantly, continue to show how data and concepts from cellular neurobiology, experimental psychology, ethology, and cognitive science may illuminate long-familiar clinical phenomena and observations.

More specifically, David Cooper traces the explicit and implicit role associating learning concepts have had in the theories and procedural recommendations of analysts from Freud onward.

371

Cooper develops the proposition that current research into classical and operant conditioning has yielded findings that mesh elegantly with and help to explicate the rationale of numerous facets of "classical" psychoanalytic technique.

In her contribution, June Hadley reexamines traditional Freudian motivational theory in light of current neuroscientific data and suggests a reformulation of familiar and perhaps overly broad concepts of sexuality. Furthermore, Hadley raises the possibility that environment-dependent learning—and not genetically programmed biologic epigenesis—may underlie aspects of "psychosexual" development and conflict.

Drawing on information theory as well as neurobiology and cognitive science, David Olds advances the idea that consciousness as a neural system receives processed sensory and affective data and serves as an evolutionarily advanced and valuable source of "read out" about what the brain and its "owner" are "up to."

Relying on concepts and findings from neuroscience, experimental psychology, and ethology, Schwartz suggests that nonverbal signs of affect can give the clinician valuable cues and guides to content and timing of interpretations.

Finally, from his vantage point as an experimentalist, Bruce Wexler reviews his own and others' research into nonconscious reception and processing of emotionally charged perceptions—work that bears on expanding our understanding of aspects of defensive operations and unconscious functioning.

In sum, this issue's five articles approach a reasonable broad sample of matters long of central concern to psychoanalysis and related disciplines—the areas of the nature of consciousness, unconscious mental functioning and perception, the neurobiology of motivation, the mechanism of therapeutic effect, and technique. The editors' hope is that our authors' offerings will not only expand our views of the specific topics they treat but also demonstrate that detailed interdigitations of neuroscientific data and concepts with familiar analytic ideas and observations are both possible and useful.

REFERENCES

Freud, S. (1895). Project for a scientific psychology. *S.E.*, 1.
Kandel, E. R. (1979). Psychotherapy and the single synapse. *New England J. Med.*,
301:1029–1037.

Andrew Schwartz, M.D.
Issue Editor

Classical Psychoanalysis and Classical Conditioning: Guilt (and Other Affects) by Association

D A V I D E. M. C O O P E R, Ph.D

NEURAL FOUNDATIONS OF LEARNING are being discovered, with far-reaching ramifications for cognitive science (Kandel, 1983; Schwartz, 1987, 1988; Rescorla, 1988; Alkon, 1989). This work establishes the importance of associative learning, a phenomenon that is of much potential interest to psychoanalysts, who, after all, attempt to learn about patients via their "free associations." The neurophysiologists' work supports the ubiquity of Pavlovian or classical conditioning in the acquisition and activation of learning in various species, including humans. Psychoanalysts may welcome these developments, for if, as I will attempt to show, psychoanalytic practice is compatible with classical conditioning theory, we may be within view of a time when, as envisioned by Freud, neural correlates of psychoanalytic hypotheses can be established (for efforts in this direction, see Meyersburg & Post, 1979; Reiser, 1984; Schwartz, 1987, 1988). Whether one gravitates toward a positivistic or a hermeneutic view of the science of psychoanalysis, it seems desirable to establish that one's view of how the mind works does not violate knowledge gained from brain science of how the brain works.

Psychoanalysis and learning theory have followed parallel and

Dr. Cooper is Director of Group Psychotherapy, Chestnut Lodge; he is a Candidate, Washington Psychoanalytic Institute.

A shorter version of this paper was presented at the annual meeting of the American Psychoanalytic Association, May 1990. The author is grateful for the helpful discussions by Lynn Whisnant Reiser and Allan D. Rosenblatt.

generally nonintersecting paths of development. While there have been notable exceptions to this theoretical isolationism (e.g., French, 1933; Dollard & Miller, 1950; Wolf, 1966; Sandler & Joffe, 1968; Greenspan, 1975; Wachtel, 1977; Reiser, 1984; Schwartz, 1987), in general there has not been much interest in rapprochement from either party to this breach. Rapaport (1960) noted that psychoanalysis lacks a learning theory. I would maintain that with recent developments in the field of classical conditioning this need no longer be the case.

Not only is classical conditioning as presented in the work of Rescorla (1988) not incompatible with psychoanalysis, a similar theory of learning is implied by familiar psychoanalytic propositions. Psychoanalysts who have been turned off by what they perceive as the barrenness of stimulus-response (S-R) theory, might take another look: Contemporary views of classical conditioning stress that what gets learned in classical conditioning is not simply a pairing of stimulus and response but complex associations between and among internally and externally mediated events for the purpose of apprehending the world. While psychoanalytically oriented workers object to learning theorists' putative presumption that the human being is a *tabula rasa* upon which the environment acts randomly through S-R associations, Rescorla's view should be more palatable.

Pavlovian conditioning is not a stupid process by which the organism willy-nilly forms associations between any two stimuli that happen to co-occur. Rather, the organism is better seen as an information seeker using logical and perceptual relations among events, *along with its own preconceptions,* to form a sophisticated representation of its world [p. 154, emphasis added].

Organisms' "preconceptions" could be, in psychoanalytic terms, the individual's drive derivatives, defenses, representational world (Sandler & Joffe, 1968), or any other descriptor of "inner life" through which the individual filters perceptions.

In this paper I will seek to establish lines of concordance between classical conditioning and "classical" psychoanalysis, two fields that recognize the importance of associational processes in human mental life but that have for the most part not recognized the natural associations between the two fields.

Classical (Pavlovian) Conditioning

Pavlovian conditioning has traditionally been conceptualized as the learning of an association between a stimulus and a response via the pairing of a conditioned stimulus (CS) and an unconditioned stimulus (US), whereby the CS comes to produce a conditioned response (CR) reliably; the CR is related to the unconditioned response (UR), which is elicited by the US. In Pavlov's experiments, dogs experienced a tone (CS) paired with the presentation of meat powder (US) to which they salivated (UR). After successive pairings of tone and meat powder, the animals came to salivate (CR) in response to the tone alone. This dry, mechanistic description appears distant from the rich data of the psychoanalytic situation. Nor has instrumental learning, in which reinforcement is seen to increase the frequency of a particular response to a given stimulus, proven any more accessible to the psychoanalytic practitioner. In two recent papers, Rescorla (1987, 1988) reviews modern conceptualizations of conditioning, both classical and instrumental, and presents a view that I hope to demonstrate is relevant to psychoanalytic conceptualizations.

Rescorla (1987) reports that early views of instrumental learning stipulated that reinforcement forges a connection for the organism between a stimulus and a response immediately preceding the reinforcement. In this view, there is no necessity to assume the experimental animal possesses knowledge or anticipation of the future; it is simply reacting to environmental contingencies. Rescorla cites research evidence that the early view was overly simplistic. It appears that not only does the animal associate a stimulus with a response, but the reinforcer itself comes to be associated with the response. Studies show that once a stimulus-response connection has been established through reinforcement, if the reinforcement is "devalued", e.g., through poisoning the previously reinforcing food, or through changing the animal's motivational state, e.g., through satiation, response frequency declines. This change would not take place if the only learned association were between stimulus and response; associations also seem to be established between response and reinforcement, such that after the association between stimulus and response has been learned and reinforcement has been discontinued, reducing the value of the previously presented reinforcer reduces the frequency of the response. For example, an animal can be trained in a reinforcement paradigm to emit a particular response, say pressing a bar, to a particular cue by giving food every time the animal presses the bar in the presence of the cue. The learning of the association between bar pressing and the cue is demonstrated by the animal's consistently pressing the bar in the presence of the cue without additional administrations of reinforcement. If at this point one decreases the desirability of the reinforcer that was used to establish the learned association, either by adding a toxin to the food or by satiating the animal, the animal will press the bar less frequently in the presence of the cue. Under normal circumstances bar pressing to the cue remains active despite the withdrawal of reinforcement once it has been learned. But in this experimental paradigm, it is as if the animal presses the bar with an internalized vision of a reward, which once devalued no longer motivates the response. A human parallel can be seen in the individual who engages in a behavior, say self-deprivation of one sort or another, for the imagined approval of an idealized other. Working through the idealization (devaluing the reward) should decrease the frequency of the behavior, in this case self-deprivation.

In Pavlovian terms the response in an instrumental paradigm can serve as a CS and the reinforcer as a US. "CSs become associated with USs to the degree that they give information about those USs; when that information is taken away by delivering the US at other times, the association fails to develop" (Rescorla, 1987, p. 121). Thus laboratory animals can be trained to press a bar to obtain food pellets; however, if food is also provided at other times, the frequency of bar pressing will decrease. The appearance of food (the US) at times other than following bar pressing (CS) weakens the association between CS and US because the CS no longer gives the animal as much information about the US, which now may appear at any time.

The view just presented represents a shift away from seeing the animal as the passive recipient of conditioning from the environment toward seeing the animal as engaged in an active process of "learning of relations among events so as to allow the organism to represent its environment" (Rescorla, 1988, p. 151). In the classical conditioning paradigm, from this perspective the random contiguity of stimuli is no longer sufficient (or necessary) to produce learned associations. It is "the information that one stimulus gives about another" (Rescorla, 1988, p. 152) that determines whether an association will be formed. This stress on the importance of the informational value of stimuli provides a bridge, I think, to human learning and to the psychoanalytic world of meanings, affects, and motivation.

In Rescorla's view, not only is the animal more actively involved in learning in the classical paradigm than has often been supposed, but the content of the learning is more complex. Rescorla reports:

Associations are formed not just between the primary events psychologists present, the CS and US. For instance, each of those events also becomes associated with the context in which they are presented . . . [Also,] modern Pavlovian thinking does not envision all of this learning taking place among simple pairs of elements all treated at the same level of analysis by the organism. Rather . . . there is good reason to believe that there is a hierarchical organization in which associations among some pairs of items yield new entities that themselves can enter into further associations [1988, p. 155].

This idea that learning takes place via successively higher order associations, insofar as these associations aid the organism in forming useful representations of the environment, can account for the complexity and richness of human mental life, which is the focus of psychoanalytic attention.

We have, thus, an outline of a process of learning along classical conditioning lines. It bears mention that Pavlov also described conditions of unlearning, or extinction. "This phenomenon of a rapid and more or less smoothly progressive weakening of the reflex [response] to a conditioned stimulus which is repeated a number of times without reinforcement may appropriately be termed *experimental extinction of conditioned reflexes*" (Pavlov, 1927, p. 49). Psychoanalysis requires a theory of unlearning as well as of learning. Wachtel (1977) and Schwartz (1987) have described analytic therapy as essentially an extinction procedure whereby the analyst, through maintaining a neutral, interpretive stance, helps the patient to unlearn certain maladaptive and pathogenic associations between cognitive content and affect. We will return to this point below.

One additional aspect of classical conditioning research Rescorla (1988) referred to, which has relevance for the present discussion, is the Garcia effect (Garcia & Koelling, 1966; Bolles, 1975). The Garcia effect refers to, "The *remarkable facility* with which rats (and a number of other animals) learn about the relationship between the taste of a particular food substance and a subsequent illness" (Bolles, 1975, p. 171, emphasis added). The original study (Garcia & Koelling, 1966) demonstrated that rats could be trained to avoid drinking bright, noisy water (water, the drinking of which was accompanied by flashing lights and noise) by following such drinking by an electric shock to the feet. Following the bright, noisy water with induced illness (via Xray or toxin) did not diminish the subsequent drinking. In contrast, avoidance of saccharin-flavored water could be trained with induced illness but not with electric shock. Garcia and Koelling speculate, "natural selection may have favored mechanisms which associate gustatory and olfactory cues [in this case saccharin] with internal discomfort since the chemical receptors sample the materials soon to be incorporated into the internal environment" (1966,

p. 124). The "remarkable facility" may well have such adaptive significance, but in any event it is a further demonstration that not just any random pairing of stimuli will produce a learned association. The Garcia effect suggests another factor that must be taken into consideration to understand learning, namely, the species-specific (or perhaps even organism-specific) neural organization of the subject that will favor the formation of certain associations. The importance of inborn factors should not be underestimated in the formation of associations. Lichtenberg (1989) reviews research on the tracking of contingency effects by infants, which he sees as fundamental to the development of the exploratory-assertive motivational system. The early ability to track contingencies is necessary for the infant's experience of pleasure in efficacy, an important element in early learning. Thus again we see the impact of the organism's "wiring."

Learning can be studied both from the perspective of teaching (Pavlov "teaching" his dogs to salivate to a tone, Garcia "teaching" his rats to avoid water), and from the perspective of the subject's memory. In clinical work, we are clearly interested in both perspectives; we endeavor to understand what was the teaching that facilitated our patients' learning what they did (e.g., to fear intimacy, to seek out danger, or to believe that they have irreparable flaws), as well as to understand the process of activation of memory (e.g., what reminds them of scenes of abuse or triumph or, on the other hand, what renders them unable to connect current experience with any past experience). Theories of memory provide us with maps to understand how individuals organize their learning. According to Reiser (1984, p. 128),

> Freud postulated a developing cognitive capacity to respond to inner and/or outer stimuli as constituting danger situations, that is, situations with the potential to develop into 'traumatic situations'. . . . Freud postulated further the laying down during mental development of networks of memory traces— residual, enduring engrams of actually experienced or fantasized events connected in time with conflicted situations and

hence with the potential to generate the same kinds of affective responses. The signaling 'system,' then, could be activated from inside by inner needs or from outside by meaningful life situations — situations that have sufficient analogous or homologous meaningful resemblance to those important earlier unresolved conflictual situations which had left traces in the nodal memory network.

Contemporary research indicates that affects attached to memories play an especially important role in the encoding and recall of associations. Bower (1981) describes the phenomenon of "state-dependent remembering," demonstrated by subjects' propensity to remember material with similar affective coloring to that of their current state of mind. Although the cognitive psychologist's laboratory may seem alien to most psychoanalysts in their consulting rooms, the phenomenon of state-dependent remembering is exactly what we observe when a patient, under the influence of a particular transference, recalls material deriving from experience with the original object(s). Lichtenberg (1989) notes, "Longitudinally, affects provide the principal thread linking infants' early presymbolic experiences, alone and with others, to later, more cognitively organized experiences" (p. 260).

Before making more explicit connections with psychoanalytic theory and practice, by way of summary I would like to indicate how the above material on classically conditioned associational learning can be applied to human learning. In the course of development humans form a multitude of associations of increasing complexity, including associations between and among internal and external stimuli, responses emitted, and those important classes of internal or external stimuli that constitute reinforcement or punishment. These associations will carry an affective component, reflecting the quality of the experience. Affect may serve as stimulus (when I feel angry [CS], I withdraw [CR]), response (when I am with others [CS], I feel ashamed [CR]), or reinforcement (doing things for others [CR] gives me pleasure [reinforcer]). Through this associational process, the individual

comes to represent and understand the world, including images of
self, others, and expectations of the environment. Memories with
similar affective qualities will tend to be associatively linked with
each other, although there may also be other important avenues of
associative linkage, e.g., according to elements of content. Associations are unlikely to change unless they are tested and found not
to be useful in understanding the world, thereby necessitating
some alternative associative learning. With the variety of
maladaptive associations we see in the clinical situation, the
treatment task becomes to facilitate the extinction of learned
associations and the learning of new associations. In the next
section, I hope to show how psychoanalytic theory and so-called
"classical psychoanalytic" treatment are consistent with principles
of classical conditioning.

"Classical" Psychoanalysis

For the purposes of this discussion, I will consider the adjective
"classical," as applied to psychoanalysis, to describe a particular
treatment approach, without necessary reference to a specific
metapsychology, because, while some elements of metapsychology may conflict with classical conditioning principles, I hope
to demonstrate that classical technique does not. First, let us
attempt to define the hallmarks of classical technique. I see
classical technique as defined by the analyst's adherence to certain
role requirements for the explicit purpose of facilitating the
emergence of and subsequent understanding of all facets of the
patient's idiosyncratic thoughts, feelings, and behaviors. Stone
(1961) has defined "the classical analytic situation," as including:

> the total or relative nonvisibility of the analyst during the hours;
> the relative confinement of his responses to interpretation,
> clarification, or other "neutral" maneuvers; the stereotypy of
> schedules and fees; the relative lack of even conventional
> emotional responses to the patient's personality and career; the

lack of intervention in the patient's everyday life, whether through advice or persuasion, or purposive extra-analytic contact; the general "blanketing" of the analyst's personality, actively and passively, except as it appears inevitably or inadvertently [p. 20].

He goes on to stress that:

I do not overlook the security and support, the high tolerance, and the sense of expressive freedom which are implicit for the patient in the psychoanalytic situation as such. However, I must state my conviction that a nuance of the analyst's attitude can determine the difference between a lonely vacuum and a controlled but warm human situation, which does indeed offer these gratifications, along with its undoubted rigors [pp. 21–22].

The classical analytic attitude, which should contribute to the creation of "a controlled but warm human situation" is described by Brenner throughout his writings. Brenner writes of the importance of "an analyst's attitude of interest in the determinants of a patient's thoughts and behavior. Whatever one says or does as an analyst should be subordinated as much as possible to that attitude or be a consequence of it" (1976, p. 31). One might add that this attitude of interest on the analyst's part may serve as an affective cue that encourages (or reinforces) the patient to proceed with analytic exploration.

Proponents of classical technique see themselves as Freudian analysts, despite the obvious fact that Freud himself did not employ what today would be considered classical technique (Lipton, 1977, 1979, 1983). I believe that Lipton's objections need not deter us from building on the genius of Freud and the accumulated experience of 50 years since his death to arrive at the best technique for expanding upon Freud's discoveries. In a footnote to a 1981 paper, Stone indicates:

I would not think it right to eat with a patient, or to raise money for him, or to get involved in physical examination or treatment

(short of an unequivocally compelling emergency). Strangely
enough I would never even think it proper to eat (or take coffee)
during a patient's hour—a "rigidity" which would make me
seem more strict than some esteemed colleagues of a few
decades ago. In mentioning the meal and the fund-raising, I am
of course indicating that I do not believe that all of Freud's
pioneering behaviors are to be accepted as precedents. But in
rejecting them (as he would have done), I would stick to the fact
that we must take extreme care lest we throw out a healthy and
promising baby with the bath! The core conception of a natural,
friendly, appropriate adult relation to another adult (the patient) within the prescribed professional limits, remains, I
believe, of decisive importance [pp. 106].

In the service of saving the "baby," let me review a few formulations of Freud's, from across the corpus of his work, that are
consistent with my attempt to reconcile psychoanalysis with
learning theory.

In his chapter on "The Psychotherapy of Hysteria" (Breuer &
Freud, 1895), Freud describes the "stratification of the pathogenic
psychical material," including:

an arrangement according to thought-content, the linkage made
by a logical thread which reaches as far as the nucleus and tends
to take an irregular and twisting path, different in every
case. . . . the course of the logical chain would have to be
indicated by a broken line which would pass along the most
roundabout paths from the surface to the deepest layers and
back, and yet would in general advance from the periphery to
the central nucleus, touching at every intermediate halting place
[p. 289].

This implicit model of associative learning in Freud's thinking
becomes explicit and is related to Pavlovian principles ten years
later in Freud's discussion of comic pleasure, where he talks about
anticipation with "expectant ideas" as setting the stage for comic

release when an occurrence does not match one's expectations. He relates this to Pavlov's work with dogs' salivation in relation to their expectation of being fed (1905, pp. 197–198).

As Freud's work with the psychoanalytic method progressed, he became increasingly aware that what is curative for a patient is not simply the recall of repressed ideational content from the associational chain but the achievement of an affectively convincing understanding via working through the patient's resistances. Simply telling a patient what is being avoided is useless and does not undo the maladaptive learning. "If knowledge about the unconscious were as important for the patient as people inexperienced in psycho-analysis imagine, listening to lectures or reading books would be enough to cure him. Such measures, however, have as much influence on the symptoms of nervous illness as a distribution of menu-cards in a time of famine has upon hunger" (1910, p. 225).

Near the end of his life, Freud employed a similarly dismissive metaphor in discussing Rank's efforts to radically shorten psychoanalytic treatment by going right for the birth trauma, which presumably lay at the core of patients' neuroses. Freud compared this approach to the situation that would arise "if the fire-brigade, called to deal with a house that had been set on fire by an overturned oil-lamp, contented themselves with removing the lamp from the room in which the blaze had started" (1937, pp. 216–217). The clinical issue here is *the necessity to work one's way through the associative links that have resulted in the patient's current state.* If the patient's situation is seen to reflect a complex series of learned associations, which are often mutually reinforcing, the question of why analysis takes as long as it does (Brenner, 1987) looks potentially answerable; the extended time required to conduct a relatively complete analysis may be a function of built-in limitations on humans' capacity to unlearn overlearned associations.

Another issue raised by Freud, relevant to the current discussion, is the revival of learned patterns of relating in the transference. "We soon perceive that the transference is itself only a piece of repetition, and that the repetition is a transference of the forgotten past not only on to the doctor but also on to all the other aspects of the current situation" (1914, p. 151). This importance of the context of the revival of associations parallels discussions across the history of classical conditioning (Pavlov, 1927; Rescorla, 1988) of the importance of context in associative learning; associations are learned not just between stimuli that are presented by the experimenter, but between any or all stimuli and the context of presentation as well (e.g., associations to the experimental apparatus, or, in the case of human learning to aspects of the environment). All aspects of this learning are potentially revivable in the transference, which applies not just to the relationship with the analyst but to all aspects of the analytic situation.

What Freud calls the "compulsion to repeat," which is admitted into the transference and "allowed to expand in almost complete freedom," (1914, p. 154), can be conceptualized as constituted by the classically conditioned associative links that get revived in the contemporary transference situation. While the clinical manifestation may look like a compulsion to repeat, ascribing such a mechanism to learned behavior may be the equivalent of saying Pavlov's dogs had a compulsion to salivate. The idea that the revival of associations in the analytic situation manifests with the force of a compulsion receives clear expression in "Analysis Terminable and Interminable."

> The adult's ego, with its increased strength, continues to defend itself against dangers which no longer exist in reality; indeed, it finds itself compelled to seek out those situations in reality which can serve as an approximate substitute for the original danger, so as to be able to justify, in relation to them, its maintaining its habitual modes of reaction. . . . The essential point is that the patient repeats these modes of reaction during the work of analysis as well, that he produces them before our eyes, as it were [1937, p. 238].

Rather than seeing "the ego" as "compelled" to seek reality justifications for its reactions, we might say, from the perspective herein, that individuals seek out situations that are consistent with expectations based on earlier learned associations and even contribute to creating situations in their interpersonal environments that reinforce their views of the world (Wolf, 1966). "Given who we are, we select and create a particular kind of interpersonal world; and given that world, we experience the need to go on as we have—and thus elicit that same kind of personal world again" (Wachtel, 1980, p. 70).

This cursory review of some of Freud's ideas has led us to the centrality of transference in psychoanalytic treatment. From the point of view of learning theory, transference can be thought of as an example of generalization (Wolf, 1966), the tendency to respond to the new stimulus situation with responses acquired in prior learning, as learned expectancies from the patient's past are activated in the psychoanalytic situation. Transference refers to the ubiquitous process whereby prior interpersonal learning affects one's perceptions and expectations in a contemporary interpersonal relationship. In his discussion of transference love (1915), Freud notes, "It is true that the love consists of new editions of old traits and that it repeats infantile reactions. *But this is the essential character of every state of being in love*" (p. 168, emphasis added). Thus, for Freud, transference was an element of interpersonal relationships, not restricted to the psychoanalytic situation. In Brenner's view, "What distinguish an analytic relationship from any other are not the *dynamics* of the transference but its *place* in the relationship, i.e., the analyst's attitude toward the transference and the use he makes of it" (1976, p. 112). The classical psychoanalytic situation, described in the quotation from Stone above, is calculated to encourage the relatively uncontaminated emergence and resolution of conflicts from the patient's past, importantly though by no means exclusively, in the transference relationship with the analyst.

Wachtel (1980) uses the Piagetian construct of schemas to explicate the phenomenon of transference. In Piagetian theory, individuals are seen as engaging in processes of assimilation, whereby aspects of the external world are apprehended through internally organized templates (schemas) which are at least partly learned, and accommodation, whereby schemas are altered to fit new experiences of the world. These concepts lead to a view of the individual as actively engaged in the process of learning, consistent with the view that emerges from modern classical conditioning theory cited above. When, in the analytic situation, transference appears especially prominent, "The experience with the analyst is assimilated to schemas shaped by earlier experiences, and there is very little accommodation to the actualities of the present situation which make it different from the former experience" (Wachtel, 1980, p. 63). Wachtel points out that accommodation of interpersonal schemas may be harder to achieve than of those relating to the physical world because of the inherent ambiguity of many interpersonal stimuli. In contrast to the physical world, where a faulty schema may be dramatically disconfirmed, thus forcing accommodation (e.g., electric stove burners that are bright orange are not the same as the child's bright orange frisbee), the ambiguity of interpersonal cues may make disconfirmation and subsequent accommodation of earlier acquired schemas harder to achieve. The purposeful ambiguity of the psychoanalytic situation should, therefore, facilitate the patient's expression of attempts to assimilate, or, in classical conditioning terms, the patient's generalization from earlier experiences, or, finally, in more traditional psychoanalytic terms, the patient's expression of transference reactions.

The discussion of transference has been intended to show that this critical construct in psychoanalytic theory and treatment is consistent with formulations derived from contemporary understanding of Pavlovian conditioning. At this point, we can begin to make more explicit some converging lines of theory between classical conditioning and classical psychoanalysis.

Association and Affects in Pathogenesis and Treatment

At the start of this paper, I related the perceived barrenness of learning theory to the view of the human organism as a *tabula rasa* which the environment shapes in essentially random fashion. Accumulated clinical experience and our own subjective experience tell us that behavior is not random, and many of us believe that a coherent general psychology must include a better theory of motivation than one that relies solely on random environmental exposures. Certainly Rescorla's work allows for the centrality of the individual's active attempts to organize input from the environment. Garcia's rats, for example, were not so malleable as to learn to associate equally all classes of paired stimuli. Humans, too, have reactions, some inborn and some learned, that make some associations more likely to be learned than others. These reactions are reflected in affects, "the consciously experienced aspect of feedback evaluative or appraisal processes, operating within a motivational system by comparing input with the goals of the motivational system. . . . It should be emphasized here that, in this conception, affect is *not* a response to a prior cognitive appraisal. It *is* the appraisal, or, at least, the conscious phase of such appraisal" (Rosenblatt, 1985, pp. 88–89).

The appearance of indicators of discrete categories of affect in newborns (Stern, 1985) supports the view that an important part of affect expression (and probably of subjective affect experience) is innate, and these categories are more specific than the broad categories of pleasure and unpleasure postulated by Brenner (1982) and others as the earliest differentiating factor among affects. Contrary to the frequently implied idea that Freud's attention to clinically relevant affects gave exclusive priority to anxiety is his disclaimer that anxiety is exceptional among the affects (1926). Regarding other affects, he states, "I should be inclined to regard them as *universal, typical and innate* hysterical attacks" (p. 133, emphasis added).

From both within (Schwartz, 1987) and without (Tomkins, 1981) psychoanalysis has come the idea that affects are not only products of an appraisal process, but that they provide motivation, whether positive or negative. Lichtenberg (1989), in his postulation of five motivational systems, sees affects as, "providing experiential targets for motivational aims" (p. 6). Based on accumulated clinical and observational research, Lichtenberg proposes that human motivation can be encompassed by "systems" in the areas of regulation of physiological requirements, attachment-affiliation, exploration-assertion, aversion, and sensuality-sexuality. Affects, as the experiential targets, are centrally involved in this schema as one is motivated to pursue, for example, the pleasure of satiation, the quiet pleasure of intimacy, the thrill of mastery, the relief of escape from unpleasure, and/or the pleasure of sexual excitement. Failures within any of the five systems will be associated with unpleasure, which will call into play the aversive motivational system, by which the individual will seek to escape from or "defend against" the unpleasure. Motivational systems do not operate alone, and while one system may predominate, the complexity of motivation is such that individuals may be influenced by many of the systems simultaneously.

Motivational systems, such as those proposed by Lichtenberg, reflect the inborn "wiring" that the individual brings to the learning task. Given the predilections and limitations imposed by these systems, associative learning, according to principles reviewed above, is the "glue" with which the individual puts together, "neurophysiologic 'Lego blocks' " (Schwartz, 1988, p. 357), to create the idiosyncratic combination of learning that we identify as "character." In the clinical situation, we may identify such maladaptive resulting complex associations as, for example, an association of assertion with aggression (Stechler & Halton, 1987). The complex association may be seen as "a compromise formation in which drive derivative, anxiety or depressive affect, or both, associated with the calamities of childhood [object loss, loss of love and castration], defense, and superego manifestations all play a role . . ." (Brenner, 1982, p. 143). In a similar vein, Freud (1926) refers to symptom formation in obsessional neurosis: "The symptom-formation scores a triumph if it succeeds in combining the prohibition with satisfaction so that what was originally a defensive command or prohibition acquires the significance of a satisfaction as well; and in order to achieve this end *it will often make use of the most ingenious associative paths*" (p. 112, emphasis added).

If we picture psychopathology as the result of learned associations, it is reasonable to see a rational treatment approach as involving, "an intricate process of unlearning—a systematic extinguishing of acquired links between cognitive content and affect that involves the examination, reappraisal, and ultimate disregarding of all congruent experiences which have forged these connections and shaped emotional response" (Schwartz, 1987, p. 493). In agreement with Schwartz, I would maintain that this is just what is provided in the "classical" psychoanalytic approach. In Wachtel's (1977) words, with parenthetical additions of my own:

> The therapist creates an atmosphere and sets up a situation in which the patient is encouraged and facilitated in talking about things that arouse anxiety [or, we might add, some other dysphoric affect]. He begins to talk about these things very tentatively and haltingly, anticipating some kind of negative or punishing response. When the therapist responds in an accepting manner, showing little or no discomfort and encouraging the patient to continue [and, we hope, interpreting the patient's associative linkages, thus rendering what had been unspeakable now speakable and understandable], some of the patient's anxiety is extinguished.
> This extinction generalizes to related thoughts and reactions, so that a related thought previously just slightly too anxiety-provoking to be free of inhibition can now occur [p. 86].

Deviations from the classical psychoanalytic position of abstinence and neutrality, or from steadfast adherence to the psychoanalytic attitude of benign interest in understanding the patient, may subtly, or not so subtly, reinforce the very learning we are trying to help the patient unlearn (Schwartz, 1990). As pointed out by Fenichel (1941), using an extreme, but perhaps not uncommon example: "When, for example, we are striving to demonstrate to the patient in their true fashion, certain character traits of his which serve as resistance, and for this purpose we imitate him, we are likely thereby to injure his narcissism. If then he becomes angry, we have not thus 'liberated his negative transference', but have simply made him angry" (p. 32). This could certainly contribute to a confirmation for the patient of fears about interpersonal contacts that therapy should remove and not reinforce.

Brenner, in his advocacy of a consistent analytic approach, suggests that the analyst forgo surplus social amenities in the service of rendering the determinants of the patient's transferences as intelligible as possible. Any such "non-analytic" interventions as advice-giving or discussing topics of mutual interest run the risk of muddying the waters of the analytic setting. Brenner (1976) states that the extent to which nonanalytic interventions disrupt a treatment will depend on the relevance they have to a particular patient: "If the relevance is great, the undesirable effect on analysis will be correspondingly great; if the relevance is small, the effect may not be of any practical significance. The difficulty is that one cannot always tell in advance whether the relevance of such an intervention is great or small" (p. 32). Parallels to Pavlov's (1927) efforts to eliminate extraneous stimuli that would make rational interpretation of his experimental results impossible serve to highlight the converging lines of theory. In the clinical situation, the risk of inadvertently reinforcing problematic learned associations, and thus making their understanding and ultimately their extinction more difficult or even impossible, dictates the preferability of an approach that relies on benign, neutral interpretation, the hallmark of the classical psychoanalytic treatment approach.

Conclusion

Knowledge of the principles of classical conditioning theory, as represented in the work of Rescorla (1987, 1988) can make a positive contribution to psychoanalytic conceptualizations of pathogenesis and the process of treatment, including issues of treatment technique. Human beings enter the world with certain constitutionally given predispositions, which psychoanalytic theorists have conceptualized under the rubric of drives or "motivational systems" (Lichtenberg, 1989). As is the case with learning in other species, the human's inborn propensities establish certain parameters within which learning can take place. Psychopathology develops through the acquisition of variously complex, maladaptive learned associations among mental representations, which include one or more affects (e.g., I am worthless; people are hostile and frightening; I am entitled to reparations for my suffering; if I compete I am bound to fail). In most cases, maladaptive learning is not an isolated association in the individual's personality, but is either blatantly or subtly pervasive; once established, an affective association may color all further learning. Thus by the time a mature person presents for treatment, a "core" conflict may have so affected the course of development that only such a thoroughgoing extinction procedure as that made possible by psychoanalysis can impact felicitously on the presenting problems.

Classical psychoanalytic technique is calculated to make possible the most extensive revival in the treatment of all aspects of prior learning that contribute to the contemporary clinical picture and to facilitate the patient's reappraisal of this prior learning via "insights" gained in the treatment. What is critical to this process is the analyst's neutral (nonreinforcing of prior maladaptive learning), interpretive approach to the patient's presentation. While a certain degree of outward passivity on the analyst's part may be critical to the creation of sufficient ambiguity in the psychoanalytic situation to allow for the most complete exposure and understanding of the patient's learned associations, I would like to make clear that overidealization of silence as a technical tool can lead to an aversive experience for the patient which does nothing to facilitate the extinction of learned associations. Rather, the classical analyst must strive to maintain Stone's "controlled but warm human situation" in helping the patient to unlearn in a situation that does not in reality recapitulate the patient's early pathogenic learning environment. The conceptualization of psychoanalytic treatment as a process whereby a patient is enabled to unlearn associative links between various mental contents and affective appraisals elucidates the association between classical psychoanalysis and classical conditioning; this construction brings us one step closer to establishing the neural foundations of psychoanalysis.

REFERENCES

Alkon, D. L. (1989). Memory storage and neural systems. *Scientific American* (July), pp. 42–50.
Bolles, R. C. (1975). *Learning Theory*. New York: Holt, Rinehart & Winston.
Bower, G. H. (1981). Mood and memory. *Amer. Psycholog.*, 36:129–148.
Brenner, C. (1976). *Psychoanalytic Technique and Psychic Conflict*. New York: Int. Univ. Press.
———— (1982). *The Mind in Conflict*. New York: Int. Univ. Press.
———— (1987). Working through: 1914-1984. *Psychoanal. Q.*, 56:88–108.
Breuer, J. & Freud, S. (1895). Studies on hysteria. *S.E.*, 2.
Dollard, J. & Miller, N. E. (1950). *Personality and Psychotherapy: An Analysis in Terms of Learning, Thinking, and Culture*. New York: McGraw-Hill.
Fenichel, O. (1941). *Problems of Psychoanalytic Technique*. Albany: The Psychoanalytic Quarterly.
French, T. M. (1933). Interrelations between psychoanalysis and the experimental work of Pavlov. *Amer. J. Psychiat.*, 12:1165–1203.
Freud, S. (1905). Jokes and their relation to the unconscious. *S.E.*, 8.
———— (1910). 'Wild' psycho-analysis. *S.E.*, 11.
———— (1914). Remembering, repeating and working-through (further recommendations on the technique of psychoanalysis II). *S.E.*, 12.
———— (1915). Observations on transference-love (further recommendations on the technique of psycho-analysis III). *S.E.*, 12.
———— (1926). Inhibitions, symptoms and anxiety. *S.E.*, 20.
———— (1937). Analysis terminable and interminable. *S.E.*, 23.
Garcia, J. & Koelling, R. A. (1966). Relation of cue to consequence in avoidance learning. *Psychonomic Sci.*, 4:123–124.
Greenspan, S. I. (1975). A consideration of some learning variables in the context of psychoanalytic theory: toward a psychoanalytic learning perspective. *Psycholog. Issues, Monogr. 33*. New York: Int. Univ. Press.
Kandel, E. R. (1983). From metapsychology to molecular biology: explorations into the nature of anxiety. *Amer. J. Psychiat.*, 140:1277–1293.
Lichtenberg, J. D. (1989). *Psychoanalysis and Motivation*. Hillsdale, NJ: The Analytic Press.
Lipton, S. D. (1977). The advantages of Freud's technique as shown in his analysis of the Rat Man. *Int. J. Psychoanal.*, 58:255–273.
———— (1979). An addendum to 'The advantages of Freud's technique as shown in his analysis of the Rat Man'. *Int. J. Psychoanal.*, 60:215–216.

_____ (1983). A critique of so-called standard psychoanalytic technique. *Contemp. Psychoanal.*, 19:35–46.

Meyersburg, H. A. & Post, R. M. (1979). An holistic developmental view of neural and psychological processes: a neurobiologic-psychoanalytic integration. *Brit. J. Psychiat.*, 135:139–155.

Pavlov, I. P. (1927). *Conditioned Reflexes: An Investigation of the Physiological Activity of the Cerebral Cortex.* Oxford: Oxford Univ. Press.

Rapaport, D. (1960). The structure of psychoanalytic theory: a systematizing attempt. *Psycholog. Issues, Monogr. 6.* New York: Int. Univ. Press.

Reiser, M. F. (1984). *Mind, Brain, Body: Toward a Convergence of Psychoanalysis and Neurobiology.* New York: Basic Books.

Rescorla, R. A. (1987). A Pavlovian analysis of goal-directed behavior. *Amer. Psychologist,* 42:119–129.

_____ (1988). Pavlovian conditioning: it's not what you think it is. *Amer. Psychologist,* 43:151–160.

Rosenblatt, A. D. (1985). The role of affect in cognitive psychology and psychoanalysis. *Psychoanal. Psychol.,* 2:85–97.

Sandler, J. & Joffe, W. G. (1968). Psychoanalytic psychology and learning theory. In *From Safety to Superego: Selected Papers of Joseph Sandler.* New York: Guilford Press, 1987, pp. 255–263.

Schwartz, A. (1987). Drives, affects, behavior, and learning: approaches to a psychobiology of emotion and to an integration of psychoanalytic and neurobiologic thought. *J. Amer. Psychoanal. Assn.,* 35:467–506.

_____ (1988). Reification revisited: some neurobiologically filtered views of psychic structure and conflict. *J. Amer. Psychoanal. Assn.,* 38(Suppl.):353–378.

_____ (1990). To soothe or not to soothe—or when and how: neurobiological and learning-psychological considerations of some complex clinical questions. *Psychoanal. Inquiry,* 10:554–566.

Stechler, G. & Halton, A. (1987). The emergence of assertion and aggression during infancy: a psychoanalytic systems approach. *J. Amer. Psychoanal. Assn.,* 36:821–838.

Stern, Daniel. (1985). *The Interpersonal World of the Infant.* New York: Basic Books.

Stone, L. (1961). *The Psychoanalytic Situation.* New York: Int. Univ. Press.

_____ (1981). Notes on the noninterpretative in the psychoanalytic situation and process. *J. Amer. Psychoanal. Assn.,* 29:89–118.

Tomkins, S. S. (1981). The quest for primary motives: biography and autobiography of an idea. *J. Personal. & Soc. Psychol.,* 41:306–329.

Wachtel, P. L. (1977). *Psychoanalysis and Behavior Therapy: Toward an Integration.* New York: Basic Books.

_____ (1980). Transference, schema, and assimilation: the relevance of Piaget to the psychoanalytic theory of transference. *Ann. Psychoanal.,* 8:59–76.

Wolf, E. (1966). Learning theory and psychoanalysis. *Brit. J. Med. Psychol.,* 39:1–10.

Chestnut Lodge
500 West Montgomery Ave.
Rockville, MD 20850

The Instincts Revisited

JUNE L. HADLEY, M.D.

WHEN FREUD (1915) PLACED THE INSTINCTS "on the frontier between the mental and the somatic" (p. 122) he anticipated our present endeavor to address the interface of psychoanalysis and neuroscience. His view of an instinct was that of an endogenous excitation or "stimulus applied to the mind"; he noted its constant effects as opposed to external stimuli which are "momentary" (p. 118). He made the assumption that "the nervous system is an apparatus which has the function of getting rid of the stimuli that reach it, or of reducing them to the lowest possible level; or which, if it were feasible, would maintain itself in an altogether unstimulated condition" (p. 120).

Freud later (1920), in "Beyond the Pleasure Principle," amended this view as follows: "The mental apparatus endeavours to keep the quantity of excitation present in it as low as possible *or at least to keep it constant*" (p. 9, emphasis added). He used the term "Nirvana principle" to denote that condition and saw it as distinct from the "pleasure principle." He viewed the pleasure principle as a transformation by libido or "life instinct" of the Nirvana principle, which he aligned with the "death instinct."

In "Instincts and their Vicissitudes" Freud proposes two groups of primal instincts: "the *ego,* or *self-preservative* instincts and the *sexual* instincts" (p. 124). By the time he wrote "An Outline of Psychoanalysis" in 1938, he had concluded that it was possible to

Dr. Hadley is in the private practice of adult and child psychiatry; she has written extensively about the interface of neuroscience and psychoanalysis.

distinguish "an indeterminate number of instincts" and that not only could they "change their aim (by displacement)" but also that they could "replace one another, the energy of one instinct passing over to another" (1940, p. 128). He states: "After long hesitation and vacillations we have decided to assume the existence of only two *basic* instincts, *Eros* and *the destructive instinct.*" Thus we are left with the primacy of libidinal and aggressive instincts (or drives). He felt these instincts were the prime movers of mental life and makes the sweeping statement: "This concurrent and mutually opposing action of the two basic instincts gives rise to the whole variegation of the phenomena of life" (p. 149). Freud (1940) also makes some important distinctions between "sexual" and "genital" concepts. "The former," he wrote, "is the wider concept and includes many activities that have nothing to do with the genitals." Furthermore: "Sexual life includes the function of obtaining pleasure from zones of the body—a function which is subsequently brought into the service of reproduction. *The two functions often fail to coincide completely*" (p. 152, emphasis added).

The erotogenic zones (as manifestations of partial areas of expressions of libidinal instincts) are described in the Three Essays (1905, pp. 183–184) as "a part of the skin or mucous membrane in which stimuli of a certain sort evoke a feeling of pleasure possessing a particular quality." Freud admits that there is still inadequate information about the biology of pleasure and unpleasure which limits an understanding of just which qualities of stimulation constitute erotogenicity, but he proposes that "the quality of the stimulus has more to do with producing the pleasurable feeling than has the nature of the part of the body concerned." Furthermore he notes that "any other part of the body can acquire the same susceptibility to stimulation as is possessed by the genitals and can become an erotogenic zone." In a footnote added in 1915 he further broadens the scope of erotogenicity to encompass "all parts of the body and to all the internal organs." I will review these concepts in the light of current neuroscientific data to substantiate or refute them. It is pleasantly surprising how well some of his concepts can be fitted onto modern neurobiology and to note that the areas in which he failed to grasp basic operating principles of the brain biased his otherwise astute observations.

In his Three Essays Freud speaks of component instincts such as scopophilia, exhibitionism, cruelty, the instinct for knowledge, pleasure derived from mechanical excitations (movement) and muscular activity all of which I will address neurobiologically. The centrality of primary erotogenic zones in the phases of development of pregenital mental organization (oral, anal, and phallic) will receive particular attention in these neuroscientific explorations.

In the main, one is left with the definite impression that Freud's concept of that which is sexual (or libidinal) is very close to the concept of sensual pleasure, which is finally focused on the genitals in adult life, and that all sensual pleasures are "sublimations" of a sexual instinct.

I will try to substantiate with neuroscientific data that sexuality and sensuality are distinct, that sexuality is one of several basic motivations, and that libidinal and aggressive (or assertive and aversive) motivations are the outcomes of a common exploratory, active, "stimulus-seeking" characteristic of the CNS.

In summary I will further show that all that is pleasurable or sensual is not necessarily sexual and that the basic motive force for behavior stems from internal generators in cells groups in the brain stem, transmitted through the reticular activating system which are modulated by stimuli both from the external and internal milieu.

Neurobiology of Activation

The brain, by its design, is an electrical generator. Each nerve cell is capable of generating an electrical charge across the cell membrane by virtue of selective ionic permeability of the membrane to sodium and potassium ions. These gradients are developed by selective opening and closing of ion channels under certain conditions and selective "pumping" of ions (sodium — out; potassium — in) during inactive periods.

Neural tissue is "excitable," and these dynamic properties persist as long as the organism is alive. Therefore neural tissue tends to be innately "active," not tending toward decrement in activity, as Freud had thought.

In addition there are several "pacemakers" in the brain which can generate rhythmical firing by virtue of a "leaky membrane" phenomenon (Hobson, 1988). Two such pacemakers are situated strategically in the diencephalon and have complex synaptic connections with cerebral cortex, basal ganglia, limbic systems, and brain stem (Scheibel, 1987, p. 1056). Ascending reticular projections split at the junction of the mesencephalon and diencephalon into dorsal and ventral leaves. The dorsal leaf enters the thalamus and ends on the nonspecific thalamic nuclei. The ventral leaf projects through the subthalamus and part of the dorsolateral hypothalamus to the septum and basal forebrain. The dorsal leaf with thalamic nonspecific neurons and their thalamo-cortical projections are thought to be the most likely source of cortical synchronous wave activity such as alpha and spindle waves. The ventral leaf, on the other hand, seems primarily involved with driving the cortex at faster rhythms (Beta waves) characteristic of the attentive conscious state and REM sleep.

The nucleus reticularis thalamus, a derivation of the sub-thalamus, lies along the lateral surface of each thalamus and straddles almost all connections between upper brain stem and cortex; it projects axons back on the mesensephalic and thalamic sources of these fibers. This contains a GABA-rich feedback system which can inhibit or facilitate thalamocortical communication. Also of interest is the fact that prefrontal cortical cells have connections with the reticularis thalamus via nonspecific thalamic nuclei which can serve an override function at higher cognitive levels. This mechanism may serve to direct selective focusing (attention) and subserve superogative control over body functions and pain, which we observe in hypnotic or meditative states (Scheibel, 1987).

It is already clear that these structures are integral parts of the reticular activating system of the brain stem. This system is phylogentically the oldest portion of the brain and is absolutely essential to life. It is this functional system which provides the background activation of all neural processes and controls the switching from one brain state to another for selective processing of information.

Instincts vs. Motivational Systems

We can now take a closer look at the differences between instinct "as a driving force in behavior" as Freud conceived it and the broader concept of motivational systems as several subsystems of hierarchically organized complex behaviors integrated and orchestrated by the hypothalamus and monitored by higher cortical centers. Freud's physiological view of stimuli has given us the concept of a "stimulus applied to living tissue (nervous substance) *from* the outside [and] discharged by action *to* the outside" (1915, p. 118). He contrasts this with an instinct, which he views as "a stimulus applied to the mind," but he further specifies that not all stimuli to the mind are instinctual. He distinguishes between the pupillary response to light (noninstinctual) and "dryness of the mucous membrane of the pharynx [thirst] or an irritation of the mucous membrane of the stomach" (hunger — instinctual). He concludes that: "A better term for an instinctual stimulus is a 'need.'" What does away with a need is satisfaction. "This can be attained only by an appropriate ('adequate') alteration of the internal source of stimulation" (pp. 118–119). This formulation leaves us with a dichotomy between the processing of internal and external sources of stimuli.

Other than the location of points of reference in the brain, I find no difference in the actual manner of information processing in internal and external sources of stimuli. Information is represented by complex interactions of ensembles of cells in widely distributed areas of the brain (Hadley, 1983) and assigned specificity by location of the activated cells. If we view the central nervous system as primarily an information processor rather than a stimulus reducer we are forced toward a broader conception of how behaviors are mediated and motivated by stimuli of all kinds (internal and external) in a spontaneously active neural substrate.

Stimuli from all sources enter the processing assembly (hippocampus, amygdala) for registration, comparison with formerly stored selection, as well as encoding into memory (Hadley, 1989). (There are also more diffuse processors of information at lower levels of complexity that always operate nonconsciously, but we are not here primarily concerned with these systems.) Stimuli (momentarily) act upon a preexisting brain state (waking, hunger, satiety) and are compared with preexisting information in memory stores (either innately programmed or learned) to yield a condition of match or mismatch. This process can be viewed as a continuum from match to mismatch (or vice versa), and various patterns resulting from the comparison yield affective instructions, which in turn are linked to fixed action patterns of affective experience and display. These patterns also are hedonically toned, depending on both the outcome of the comparison and the preexisting brain state. (Food does not look appealing in a state of neurophysiological satiety whether from a large meal or anorexia nervosa!) Affectively tagged goal-directed behaviors are organized at many levels in the neuraxis and coordinated at the level of the hypothalamus and basal ganglia, which represent the effector organs of internal and external output (Hadley, 1989). The hypothalamus, carrying out the affective instructions from comparator processes, coordinates the activation of component parts of complex response patterns lower in the neuraxis into an output that is organized and also goal directed (stimulus specific). A good example is Bard's "sham rage," which can be elicited at many lower levels of the neuraxis but lacks the direction, specificity, and longer-lasting effects of hypothalamically modulated true rage (Satinoff, 1987). The hypothalamic output deals particularly with the autonomic system, whereas the basal ganglia deal more with the motor system, particularly with component motor plans mediated by the extrapyramidal system, which operate at a nonconscious level. The higher cortical centers (frontal and prefrontal) coordinate control over both output systems, allowing for priority considerations and delay in behavioral response. Therefore the organism is stimulus directed but not stimulus bound and is guaranteed the flexibility of permitting activation of the output structures or of vetoing them or further choosing consciously to emit a different behavioral output (Libet et al., 1979). The routine procedures of habitual everyday life, such as driving or brushing the teeth, do remain automatic stimulus-response patterns operating at a nonconscious level. Indeed it has been demonstrated in mammals that the same hierarchical arrangement of complex response components holds for feeding, thermoregulation, sex, aggression, and all forms of motivated behavior (Satinoff, 1987). Only when these become ineffective, yielding mismatch signals, or stimuli that are novel or unduly strong enter the system does the system engage higher cortical centers for conscious processing to devise more complex contingencies.

It is tempting to speculate that Freud with his internal- versus external-stimulus paradigm was appreciating the difference between hypothalamic and basal gangliar output systems. We must be careful, however, not to overdelineate these systems because they are intimately interconnected, particularly through the nucleus accumbens, and usually operate cooperatively.

We can now see how a concept of motivational systems with their hierarchical representation in the neuraxis can expand our view of how behaviors are assembled according to task and carried out systematically but flexibly according to internal state and stimulation. We would conclude, however, that Freud's early attempt at accounting for motivation with "instincts" was admirable (even if inadequate) and in keeping with the scientific data he had available. I do, however, disagree that the "instincts" should basically apply to libidinal and aggressive action patterns. The latter class would seem to most closely fit affective output patterns that coincide, respectively, with assertive and aversive reactions to stimuli. The sexual refers to a separate motivational system.

Pleasure, Erotogenicity, Sensuality, and Sexuality

Freud defined erotongenic zones as "part of the skin or mucous membrane in which stimuli of a certain sort evoke a feeling of pleasure possessing a particular quality" (1905, p. 183) and then merged sexuality with sensual pleasure. He was aware that the state of neurophysiological data in his time left him unable to make statements about the generation of pleasure or the distinctions between sexual and sensual pleasure.

Neuroscience has now progressed to where we know quite a bit about the generation of pleasure and unpleasure. These hedonic components of affective processes are added to the registration of stimuli as a result of comparator functions in hippocampus and amygdala (Hadley, 1989).

Not all stimuli evoke pleasure or unpleasure. As a matter of fact, the majority of incoming signals render an indifferent sensation (Cabanac, 1987).

It seems that passage through the amygdala is essential to add hedonic tone as a part of affect to any stimulus event (Hadley, 1985). Most stimuli are processed simply through the hippocampus neutrally but lack the pleasure-unpleasure labeling of an amygdalar-processed event. A few stimuli, mostly chemical, thermal, and mechanical in nature, have primary positive reinforcement qualities. Sweet taste is one example of an inherently rewarding stimulus, and mammals will work for saccharine rewards (which are non-nutrative) as well as for sugar (which satisfies a biological, homeostatic need). Thermal stimuli are usually noted for pleasurable states only in *transition* from one state to another and are indifferent in stable states. Mechanical stimulation of many sorts leads to pleasure at moderate levels, but many elicit unpleasure at higher intensities (Cabanac, 1987).

Other stimuli acquire reinforcement value (positive or negative) by associative learning (Cabanac, 1987). Reinforcement value is dependent on the ability of a stimulus to elicit hedonic affective activation. In the processing of stimuli in the hippocampus of mammals, three event-related evoked potential components have been identified (Deadwyler, 1987). The first, N1, varies with novelty. The second, N2, which is dependant on intact afferents from the medial septal nucleus, signals presence or absence of reward contingency. N2 increases amplitude proportionate to the rate of acquisition of stimulus-discrimination behavior and disappears with extinction. The third wave, P2, distinguishes between positive and negative trials and is usually larger on nonrewarded trials. P2 amplitude appears to code the strict associatively conditioned features of the sensory stimulus. The three wave components behave as anatomically independent from one another, and destruction of respective afferent pathways affects only one wave. This indicates that each convergent major afferent pathway to the hippocampus transmits a different *type* of information, which can then be integrated. The fact that N2, with its input from the thalamus through the medial septal nucleus, has an "affective tag" to indicate whether or not a reward contingency was present in the past alerts the organism early to the relevance of any given stimulus and would appear to represent "affective memory" retrieval.

Mediation of pleasure depends on the intactness of the dopaminergic circuits of the ventral tegmental area and their target, the nucleus accumbens (Wise, 1983). Activity in these dopaminergic neurons appears to produce the subjective sensation of euphoria (Hadley, 1989). Opiate reward mechanisms coexist in the ventral tegmental area, and an excitatory effect on dopaminergic cells is presumed (Wise, 1983).

It appears that as information that has rewarding potential is passed through the amygdala and is further routed to the ventral tegmental area and/or nucleus accumbens largely via the hypothalamus, the subjective experience of pleasure (or euphoria) is produced by activity in the dopaminergic circuitry. This appears to be the system Freud labeled as sexual-pleasurable.

The "satisfaction" paradigm is more closely related to the tendency to repetition that underlies the repetition compulsion and maintenance of familiarity. It would seem likely that this mechanism is the substrate for Freud's conclusion that the CNS at least tries to maintain a situation of *constancy* of neural activity.

The unpleasure or punishment mechanism is mediated through the central gray area around the Aqueduct of Sylvius in the mesencephalon, extending upward into periventricular zones of the hypothalamus and thalamus. This system is a rostral extension of the somatic-pain-mediation system and uses classical excitatory neurotransmitters (glutamate or aspartate) and the neuromodulator substrate P (Terenuis, 1987). Activity in this system is triggered when passage through the amygdala has yielded a mismatch signal and calls for an aversive response (Grossman, 1987). The aversive (fight, flight) responses are organized from the lower brain stem through the hypothalamus to most limbic structures *except* the hippocampus, which shows no effects of lesioning on aggression, escape avoidance, or "freeze" behaviors.

Hippocampal stimulation produces behavioral alerting or arousal reactions and exploratory activity, but there is no clinical or experimental evidence that such stimulation has affective consequences (Grossman, 1987).

It seems quite evident that while the hippocampus serves as a comparator, it segregates neutral stimuli from affectively labeled reinforcer stimuli, which are particularly important to the self and its survival, and which then are processed by the amygdala. Extrinsic information enters both CA3 (amygdalar routed) and CA1 (mammilary-body routed) areas, but the CA3 neurons are presynaptic to CA1 afferents so as to produce an initial segregation of incoming stimuli (Olds et al., 1989). In general, pleasure signals usefulness and unpleasure signals danger, but we all know from clinical experience with addictions that this is not always true. I very much agree with Schwartz's (1990) proposal that the pleasure mechanisms serve a mood/self-esteem regulating function in the normal scheme of things. His concept of "getting high" as an antidote to the anxiety/depression arising from narcissistic injury is highly convincing. (Incidentally, "getting even" also creates a "high.") The amygdala is then the gateway to all that is affective, hedonically toned, and representative of a "self" system. After amygdalalectomy, there is a gross flattening of affect and loss of "meaning" of stimuli (Smith, 1987; Pribam, 1980).

Now that we have gained considerable understanding about the mediation of pleasure and unpleasure, we can explore some characteristics of sensory systems in general and the genital sensory system in particular. Because genital sensory receptors are in a special class, I will focus on cutaneous sensory receptors.

The primary function of sensory receptors is to convert (or transduce) natural stimuli into the standard neural code of action potentials propagated from the periphery to central areas of the brain (Iggo, 1987). These signals are proportional to the strength of the stimulus. There are three general categories or receptors: mechanoreceptors, thermoreceptors, and nocioceptors. We are most concerned with mechanoreceptors of which there are two categories: elaborate—usually encapsulated structures lying at the dermoepidermal border or slightly buried in the dermis—and simple unencapsulated terminals.

The encapsulated endings include Pacinian corpuscles, Meissner corpuscles, Merkle cells, and Ruffini ending, which pick up different frequencies of displacement (from 1500 HZ down to steady pressure respectively). Iggo (1987) has dramatically demonstrated that isolated activation of an individual sensory receptor can result in distinct sensory perception. Meissner corpuscles mediate touch, Merkel receptors evoke a sense of pressure. Encoding of specific information is already begun by sensory receptors in the skin, and the central nervous system makes further analysis using the elementary building blocks of each discrete stimulus.

The description of the genital sensory system that follows is based on Rose (1987). This system has a unique constellation of properties. It encompasses sensory discriminative operations, affective and motivational characteristics that are unique due to their central connections, and behavioral and endocrine-control functions vital to reproduction. Sexual dimorphism is most pronounced in the genital sensory system, yielding quite different configurations of the periphery of the body in the two sexes.

In males, the sensory innervation of the genitalia has its greatest density in the glans penis, with lesser densities in prepuce, penile shaft, and scrotal skin. Encapsulated endings are typically represented by genital corpuscles. Free endings are also present, and warm and cold thermoreceptors have also been identified. The mechanoreceptors have specific dynamics of slow or fast adaptation, small receptive fields, and high rates of displacement-elicited discharge. The effects of erection on receptor function remain obscure. Sensory inputs from the penis and surrounding structures run in pudendal nerves, which then diverge, entering the dorsal roots at several levels of spinal segments, primarily S2. From there they ascend the dorsal columns to terminate in nucleus gracilis.

In females, genital sensory innervation is most dense in external structures, including the clitoris, labia, and vaginal vestibule. The vagina is least innervated, with the cervix occupying an intermediate position. Female external genitalia have the same kinds of genital receptor types as males; however, the vagina and cervix are innervated by free endings. Mechanoreceptors in the external genitalia have very restricted receptive fields and rapidly adapting responses. Vaginal afferents exhibit tonic responses to stretching of the vagina or pressure on the cervix. Afferents from the female external genitalia run through the pudendal nerves to the cord entering predominantly with S2 as in males, but vaginal afferents ascend to the cord through hypogastric nerves entering the last two thoracic and first two lumbar segments.

In both sexes perineal musculature receive extensive innervation, some of which arise from muscle spindles (Rose, 1987).

Spinal dorsal-horn neurons with genital cutaneous receptive fields show rapidly adapting mechanoreceptive properties and have small receptive fields in the ipsilateral genital structures or extensive distributions including adjacent extragenital skin regions. Vaginally responsive neurons have a more visceral source and show long-duration responses to vaginocervical stimulation.

In subcortical regions both male and female genitalia are somatotopically represented in the nucleus gracilis of the medulla and in the lateral region of the ventroposteriolateral thalamic nuclei (VPL). Cortically the genitalia are principally represented in the first somatosensory projection zone (on the dorsomedial postcentral gyrus) and in the second somatosensory cortical region. These lemniscal neurons display rapidly adapting tactile responses. The cervix is represented in the trunk projection zone of VPL and in the first somatosensory cortical region *plus* the orbital cortex.

There is also an extensive extralemniscal system that responds to genital stimulation. Most of the work with this system has been done with females, but it is assumed to exist similarly in males. Vaginal stimulation in the female evokes responses in neurons that are distributed widely in the brain stem reticular formation, in some cranial nerve motor nuclei, the central gray, and the midbrain tectum. These brain stem genital sensory neurons share certain unique properties: (1) rapid change in firing, which is enhanced by stimulus repetition; (2) stronger responses to cervical than vaginal stimulation; (3) long poststimulus duration; (4) diverse response patterns; (5) convergent responsiveness to innocuous or nociceptive mechanical stimulation of visceral and extragenital cutaneous regions. Vaginal stimulation produces central activity in the medial and lateral hypothalamus, the subthalamus, medial intralaminar, and posterior thalamic nuclei and in limbic structures including septal and amygdaloid nuclei and the cingulate gyrus. Responses observed in limbic and hypothalamic neurons are of longer latency and of simpler pattern than those observed in brain stem or thalamic genital sensory neurons.

The unique combination of lemniscal and extralemniscal components of genital sensory representation give this system its particular quality. Extra lemniscal components are considered the generators of motivational and affective properties of genital sensations. The unique attributes of this system are the result of their specialized response dynamics, cross-modal interactions between genital and extragenital somatic systems, and the intrinsic properties of the anatomical substrates involved. The extralemniscal system is also important in the neuroendocrine mechanisms regulating fertility such as induced ovulation and prolactin release. An additional interesting observation is that neural responsiveness to cutaneous tactile sensitivity is facilitated by the action of estrogen. Also of interest is the observation that vaginal stimulation in humans produces a pain attenuating effect.

With this extensive data now available, it is possible to state unequivocally that the sensory systems in general are distinct from the genital sensory system, which has unique properties and which has extensive extralemniscal representation. *Affective, hedonic responses can elicit pleasure directly through processing through the amygdalar system, but the sensations do not have the quality of erotogenesis unless they are mediated by the genital sensory system.* This is not to say that any object or part of the body cannot be "erotized" by associative learning of genital activation simultaneously with other sensory systems. The representations of a stimulus event could include both genital and nongenital components by association, so that if the nongenital stimulus is aroused a genital stimulus component will also be aroused from memory. However, the latter representation can be acquired only through learning and the development of complex representations (Hadley, in preparation). Therefore, purely sensual systems may overlap with the genital sensual system in sexual activity but may be fairly distinct from them also in the pursuit of pleasure. Sexual pleasure is a distinct form of pleasure and all that is pleasurable is definitely not sexual. We therefore need not view all pleasure as sublimated sexual pleasure.

The Erogenous Zones and Libidinal Stages of Development Revisited

When Freud postulated oral, anal, and phallic periods of development in the Three Essays, he based his theory primarily on erotogenic zones. He stated:

The character of erotogenicity can be attached to some parts of the body in a particularly marked way. There are predestined erotogenic zones, as is shown by the example of sucking. The same example, however, also shows us that any other part of the skin or mucous membrane can take over the functions of an erotogenic zone, and must therefore have some aptitude in that direction. Thus the *quality of the stimulus has more to do with producing the pleasurable feeling than has the nature of the part of the body concerned* [p. 183, emphasis added].

He seems, however, to be placing a disclaimer on the zonal theory in the last sentence and we again get the confusion of "pleasurable" and "sexual."

To understand what is possible in terms of neural functioning we must review the recent finding of Chugani and Phelps (1986), using PET scanning with radioactive deoxyglucose to follow the maturational sequence in the developing brain. At birth glucose metabolism (which signifies functionality) is highest in the sensorimotor cortex, thalamus, mid-brain, brain stem, and cerebellum, particularly the vermis. The rate of metabolism is very low in the basal ganglia and the remaining cortex. By the third month the basal ganglia have reached a functional level equal to that in the thalamus, and there is increasing activity in the transverse temporal gyrus (which includes both primary and association sensory areas) presumably including the hippocampus. The amygdala is too small a structure to get reliable resolution with current techniques. By eight to nine months the PET images show increased glucose metabolism in frontal and association cortices. This progressive increase in glucose utilization is consistent with anatomical studies showing expansion of dendritic fields and an increase in capillary density, particularly in the frontal cortices. The cingulate cortex is inactive at birth, but shows gradual increase in functionality through the first year. This functional situation is felt to coincide with cognitive or hypothesis-forming development in the infant.

The functional capacities of these areas of the brain can inform us as to what is possible in terms of experience in the infant. At birth the basic sensorimotor structures are active and monitor all sensory input from the periphery with the somatic sensory sensations being earliest to mature, the auditory systems next, and the visual systems last. Thus the neonate has much sensory input which is registered in short-term memory as opposed to "retentive" or "memory" *of* learning, which becomes available with beginning of hippocampo-amygdalar processing. This latter development marks the beginning of comparator functioning and the possibility of the addition of affect and "pleasure" to experience. This maturation is signaled by the appearance of the social smile between three and six weeks when a high-pitched human voice is the most potent elicitor (Wolff, 1963). This smile is quite different from earlier reflexive forms in that the eyes are bright and focused. Prior to this development, the infant has a group of spontaneous behaviors not elicited by external stimuli but emitted mainly during sleep or REM states, including startles, reflex smiles, erections, and sucking movements which are attributed to spontaneous neural discharges from the mid-brain stem (Stone et al., 1973), where these reflexive behaviors are organized.

Considering these data together, we can propose an early neonatal condition in which stimuli are recorded as sensual events along a comfort-discomfort gradient but do not yet have the discrete properties of affects or true pleasure-unpleasure tagging. The condition of interest due to the engagement of orienting reflexes is present at birth, and facial expressions of pain to noxious stimulus and reactions to sweet, bitter, and sour taste are innately preprogrammed. Ninety percent of two-month-olds can show anger. The sadness display is observed fully by three to four months, but fear is not reliably identified as a response to extreme novelty until seven months (Izard & Malatesta, 1987). The gradual unfolding of these affects suggests that the infant is maturing in cognitive processing in such a way as to perceive more discrete outcomes of the matching process with the consequent addition of specific affective patterning. Therefore we can conclude that the neonate has a preaffective stage of development before the functionality of the true pleasure response and that the reflex activity of sucking exists as part of an inborn action pattern which can operate even prenatally without postulating any significant engagement of limbic structures not yet matured. Later elaborations of sucking behaviors become pleasurable largely by their power to soothe (decrease arousal levels in the reticular activating system, and decrease heart rate) (Birns et al., 1973). However, sucking has been shown to be no more effective in soothing than rocking, warmth (particularly applied to the feet), and human voices (particularly high-pitched ones) (Birns et al., 1973).

These data coupled with absence of genital sensory receptors in the mouth would tend to rule the mouth out as a primary erogenous zone. Its very dense receptor representation makes it a highly "sensitive" area and can lead to sensual pleasure from several modalities (touch, taste). Also, the area can be "erotized" by learning, as it is in many cultures in the kiss, for example, but this is not obligatory.

The act of nursing is both sensually pleasurable and erotic for the mother, however, and apparently there are genital sensory representations in the erectile tissue of the nipple (a topic much neglected in the literature). In this century we emphasize the nursing event as a symbol of the establishment of trust rather than being analogous to sexual passion, as Havelock Ellis proposed in 1900. Kagan (1984) sums it up nicely: "Contemporary theorists award sex and hunger less importance and regard the individual's needs for a trusting, loving relationship and for control of anxiety as more potent forces in human behavior" (p. 85). The reliability with which an infant can shape his reflexive sucking behavior to the attainment of both nutrition and soothing lays the earliest representations of mastery in the nervous system and leads to an expectancy of success. Failures at this early phase tend to set the stage for failure at later epoches by leaving memory traces of mismatch conditions.

It is clear from my earlier descriptions of the innervation of the anal, urethral, bladder, testicles, and scrotal and surrounding skin areas with genital system receptors that any activity in these areas has the capacity to produce genital-sexual sensations after this system matures. It would appear, however, that full genital responsivity with masturbatory excitability does not develop until about 18 months. An excellent study of the stages from sensitivity to sensuality with pleasure to genital self-stimulation with self-absorption and mounting excitement is presented by Kleeman (1965, 1974). Kleeman (1965) believes that in the first year, "genital tactile self-stimulation and visual exploratory behavior had the *primary* aim of establishing a closer relationship with his body and the erotic aim was distinctly secondary, in the sense that intentional self-arousal and self-absorption qualities were not prominent" (p. 249). This observation leads us to conclude that whereas pleasure mechanisms become available between three and six weeks of age, the genital sensory system is not fully functional until about 18 months of age. It is generally agreed among infant observers that the period between 18 and 24 months for both boys and girls is one of increasing genital and perineal awareness, greater sphincter control and increasing mobility (Lichtenberg, 1989). In addition, the acquisition of language at about the same time leads to the ability to cognize symbolically. All of these newly acquired competencies lead to a vitalized surge of autonomy which inevitably produces conflicts with significant others in the environment. We would view these developments as the cause of the particular constellation of behaviors in the so-called "anal" period, namely negativism and oppositionalism.

There are significant differences between the male and female brain and behavior from intrauterine life onward. (For a detailed discussion of masculinization of the brain, see Hadley, 1989.) Sexual identity is fairly well solidified by 18 months of age, and masculinity and femininity of gesture, movement, interactional patterns, activation levels, and sensory processing are quite distinct in most cases.

The cingulate cortex is particularly important in mediating not only maternal behaviors but also play-behaviors essential to group affiliation and separation distress (MacLean, 1986, 1987), as well as separation calls and "separation feeling." This area is rich in opiate receptors and is almost surely an important area in infant-maternal bonding (MacLean, 1986). In primates it appears that these behaviors among peers are the absolute requirement for later adult heterosexual behavior and can substitute for absent, brutal, or surrogate mothering, but rearing with a good mother cannot substitute for the absence of peers (Harlow in the Competent Infant, 1973). These findings parallel findings in isolated premature and institutionalized babies (Stone et al., 1973).

In their third year with their growing complement of capacities, both boys and girls react to one another with typically masculine and feminine behaviors (Lichtenberg, 1989). It is probable that the maternal behavior seen in girls at that age is analogous to the boys' "rough and tumble" phallic play, which would suggest that the cingulate cortex has become fully functional. Paternal behavior is probably also encoded now (MacLean, 1986).

Beginning at about 18 months and progressing until age three with the growing urge toward autonomy, the infant shifts from primarily dyadic to triadic relationships, which introduces the possibility of competition and/or cooperation. By about three most infants prefer members of the opposite sex, which may be related to cingulate maturation bringing in stereotyped behavior programs appropriate to each sex. The development of full-fledged oedipal triads appears to be less complete in American culture today because of heavy work demands on both parents, separation, divorce, and the trend toward day care and the many contacts with other children and adults which dilute the child's parental attachments. When *intense* dyadic relationships with erotic overtones form with one or the other parent they usually reflect a highly unsatisfactory marital situation in which the child has become a primary significant other for the parent and has been overstimulated. Instead of aberrations of the oedipal constellation we are currently seeing more of the effects of early childhood sexual abuse with its premature arousal of erotic sensations (Hartman & Burgess, 1989). The inevitable distortions to the entire developing organism serve as one of the major pathogens of nonorganic mental illness. These sorts of experiences, particularly in the two-to-three-year age range, severely damage trust and turn the child inward toward isolation. Such experiences also produce dissociated memory traces, which cannot be accessed evocatively but remain nonconscious and partially active due to the lack of an appropriate affective tagging that would allow integration with self-representation and closure. There exists a brain mechanism that separately encodes the "affective" dimension of memory without at the same time encoding its content (Gloor, 1986). Perceptual and mnemonic phenomena are attributed to the activation of the mechanisms in the temporal neocortex (hippocampus), whereas affective states are attributable to activity in the amygdala (Gloor, 1986). The mechanisms for switching into altered states lies in the prefrontal/nucleus reticularis thalamus connection which presumably can alter the efficiency of thalamocortical information flow. I hypothesize that this is the mechanism engaged in the dissociation regularly reported by abused individuals (Benson, 1987). We must conclude that many mental contents once relegated to the status of fantasy are now observable, with systematic regression, to be subjective experiential memories lacking an appropriate affective component, which must be reactivated by anamnesis and integrated into the self representation. The process of dissociation places a "stop" on the information processing before affect can be experienced, presumably to protect the vulnerable organism from disorganization.

Neuroscientific data now available, then, make it possible to draw the following conclusion: (1) The central nervous system is inherently active rather than "powered" by instincts. Its goal is to retain familiarity and keep a hedonic balance which optimizes overall functioning. (2) Freud's division of instincts into libinal and aggressive components was a precursor of our understanding that there are assertive and aversive outcomes of information-comparator functions springing from a common exploratory source. (3) Pleasure systems are distinct from genital sensory systems, though they may overlap during development. (4) Genital sensory systems innervate genitals, perineal structures, and anal and urethral structures plus the nipples (particularly in the female) but not the oral cavity. (5) There are important extralemniscal pathways to central representations of sexuality that apparently contribute the preremptory motivational quality to genital sensory stimuli. (6) Any area of the body can be "erotized" by including it with the representation of genital sensory stimuli, by associative learning. (7) The sensual and sexual systems are distinct, but the former may serve the latter and usually does. (8) Sex differences are due not only to genetic assignment but to functional differences in the brain as well as assignment by the milieu of appropriate sex roles. (9) Masturbatory activity first serves a sensual and soothing function and only later at about 18 months becomes a truly erotic event. (10) Concurrent advances in autonomy, sphincter-control maturity, cognition, symbolic representation, and the urgency of self-assertion contribute to the constellation of behaviors usually regarded as "anal." (11) With the maturation of the cingulate cortex in the second year of life, the boy demonstrates rough "phallic" play while the girl chooses doll play and "maternal" behaviors as manifestations of sexual identity. (12) The outgrowing of the need for a tight dyad and the cognitive capacity to deal with triads accounts for the development of competition, jealousy, and cooperation. (13) Most so-called "neurotic" psychopathology is not attributable to vicissitudes of an oedipal triad, but to the noxious effects of incompetent and irresponsible parenting, which allow for traumatic overstimulation (often sexual) in infancy. (14) Infantile sexual trauma (of both males and females) accounts for a significant proportion of the nonorganic mental illnesses seen today and appears to be the most potent pathogen when coupled with totally unprotective parenting (i.e., abusive father/mother and weak compliant mother/father) operating in the sexual arena in infancy. This constellation affects all motivational systems, particularly the attachment/affiliative system and can produce almost any symptom constellation.

I would like to conclude these explorations with a recommendation that we adopt a new and neuroscientifically sound view of motivation as proposed by Lichtenberg (1989). I echo his sentiments when he states that: "Each child's innate program plus his or her parents, living environment, and cultures, are so distinctive that *individualizing* the idiosyncratic preferences, restrictions, and aversions affecting sensual-sexual motivation is as powerful an organizing principle as the view of a sequence of oral, anal, and phallic-oedipal phases" (p. 242).

REFERENCES

Benson, H. (1987). Relaxation response, physiology, history and clinical applications. In *Encyclopedia of Neuroscience.* Boston: Birkhauser, pp. 1045-1047.
Birns, B., Blank, M., & Bridger, W. H. (1973). The effectiveness of various soothing techniques on human neonates. In *The Competent Infant,* ed. L. J. Stone, H. T. Smith, & L. B. Murphy. New York: Basic Books, pp. 295-300.
Cabanac, M. (1987). Pleasure (Sensory). In *Encyclopedia of Neuroscience, 2.* Boston: Birkhauser, pp. 956-957.
Chugani, H. T. & Phelps, M. E. (1986). Maturational changes in cerebal function in infants determined by F.D.G. positron emission tomography. *Science,* 231:840-843.
Deadwyler, S. A. (1987). Evoked potentials in the hippocampus and learning. In *Encyclopedia of Neuroscience, 1.* Boston: Birkhauser, pp. 411-412.
Freud, S. (1905). Three essays on the theory of infantile sexuality. *S.E., 7.*
_____ (1915). Instincts and their vicissitudes. *S.E., 14.*
_____ (1920). Beyond the pleasure principle. *S.E., 18.*
_____ (1940). An outline of psycho-analysis. *S.E., 23.*
Gloor, P. (1986). Role of the human limbic system in perception, memory, and affect: Lessons from temporal lobe epilesy. In *The Competent Infant, ed. L. J. Stone, H. T. Smith & L. B. Murphy. New York: Basic Books, pp. 159-170.*
Grossman, S. P. (1987). Motivation, aversive, biological bases. In *Encyclopedia of Neuroscience, 2.* Boston: Birkhauser, pp. 685-688.
Hadley, J. (1983). The representational system: A bridging concept for psychoanalysis and neurophysiology. *Int. Rev. Psychoanal.,* 10:13-20.
_____ (1985). Attention, affect, and attachment. *Psychoanal. & Contemp. Thought,* 8:529-550.
_____ (1989). The neurobiology of motivational systems. In *Psychoanalysis and Motivation,* ed. J. D. Lichtenberg. New York: The Analytic Press.
Harlow, H. & Harlow, M. (1966). Learning to love. *Amer. Scientist,* 54:244-272.
Hartman, C. R. & Burgess, A. W. (1989). Sexual Abuse of Children: Causes and Consequences. In *Child Maltreatment,* ed. D. Cicchetti & V. Carlson. New York: Cambridge Univ. Press, pp. 95-128.
Hobson, J. A. (1988). *The Dreaming Brain.* New York: Basic Books.
Iggo, A. (1987). Sensory receptors, cutaneous. In *Encyclopedia of Neuroscience.* Boston: Birkhauser, pp. 1079-1080.
Izard, C. E. & Malatesta, C. Z. (1987). Perspectives on emotional developments I: Differential emotions theory of early emotional development. In *Handbook of Infant Development,* ed. J. D. Osofsky. New York: Wiley, pp. 494-554.
Kagan, J. (1984). *The Nature of the Child.* New York: Basic Books.

Kleeman, J. (1965). A boy discovers his penis. *Psychoanal. Study Child,* 20:239–265.

———— (1974). Genital self-stimulation in infant and toddler girls. In *Masturbation,* ed. I. Marcus & J. Francis. New York: Int. Univ. Press, pp. 77–106.

Libet, B., Wright, E. W., Jr., Feinstein, B., & Pearl, D. K. (1979). Subjective referral of the timing for a conscious sensory experience. *Brain,* 102:193–224.

Maclean, P. D. (1986). Culminating developments in the evolution of the limbic system: The thalamocingulate division. In *The Limbic System,* ed. B. K. Doane & K. F. Livington. New York: Raven, pp. 1–28.

———— (1987). Triune brain. In *Encyclopedia of Neuroscience.* Boston: Birkhauser, pp. 1235–1237.

Olds, J. L., Anderson, M. L., McPhie, D. L., Staten, L. D., & Alkon, D. L. (1989). Imaging of memory-specific changes in the distribution of protein kinase C in the hippocampus. *Science,* 245:866–869.

Pribam, K. M. (1980). The biology of emotions and other feelings. In *Emotion: Theory, Research, and Experience,* 1, ed. R. Plutchik & H. Kellerman. New York: Academic Press, pp. 245–249.

Provence, S., & Lipton, R. C. (1973). Infants in institutions. In *The Competent Infant,* ed. L. J. Stone, H. T. Smith, & L. B. Murphy. New York: Basic Books, pp. 795–806.

Rose, J. D. (1987). The genital sensory system. In *Encyclopedia of Neuroscience,* 1. Boston: Birkhauser, pp. 459–460.

Satinoff, E. (1987). Drives, biology of. In *Encyclopedia of Neuroscience,* 2. Boston: Birkhauser, pp. 342–345.

Scheibel, A. B. (1987). Reticular formation, brainstem. In *Encyclopedia of Neuroscience,* 2. Boston: Birkhauser, pp. 1056–1059.

Schwartz, A. (1990). On Narcissism: An(other) Introduction. In *Pleasure Beyond the Pleasure Principle: The Role of Affect in Motivation, Development, and Adaptation,* Vol. I, ed. R. A. Glick & S. Bone. New Haven, CT: Yale Univ. Press, pp. 111–137.

Smith, O. (1987). Emotion, neural substrates. In *Encyclopedia of Neuroscience, 1.* Boston: Birkhauser, pp. 385–386.

Stone, L. J., Smith, H. T., & Murphy, L. B. (ed.) (1973). *The Competent Infant.* New York: Basic Books.

Terenius, L. (1987). Pain, chemical transmitter concepts. In *Encyclopedia of Neuroscience,* 2. Boston: Birkhauser, pp. 901–903.

Wise, R. A. (1983). Brain neuroral systems mediating reward processes. In *The Neurobiology of Opiate Reward Processes,* ed. J. S. Smith & J. D. Lane. New York: Elsevier/North Holland, pp. 406–437.

3158 Booth Heights Road
East Troy, WI 53120

Consciousness: A Brain-Centered, Informational Approach

DAVID D. OLDS, M.D.

THIS PAPER IS THE SECOND in a series exploring the nature of a *brain-centered psychology*. In the first paper (Olds, 1990) I developed the idea of a psychology that would consider the brain as its central organizing concept, in lieu of previous abstractions such as ego, id, or internal objects. The rationale for this change is to organize all of our thinking about mental agencies, mental process, and psychological function in general with reference to brain processes. The goal of such a psychology, not fully realizable as yet, is to ground all mental processes in biology in a manner resembling that of Freud's early hopes. The model is an information-processing one, in which information and semiotic functions serve to bridge the gap between brain and mind.

I here pursue this idea further in discussing the phenomenon of *consciousness*. Once this model is developed I will try to show that such a model places the psychoanalytic concepts of the *unconscious* and of *psychoanalytic therapy* in a different light, bringing them into line with current research in brain function and biology. The ultimate aim of such an endeavor is to ground psychoanalytic theory in a currently relevant scientific context.

Writing about consciousness is a kind of quicksand. The topic is vast, the history of thought about it is nearly infinite, and the relevant research findings are accumulating faster than one can

Dr. Olds is Training and Supervising Analyst, The Columbia University Center for Psychoanalytic Training and Research.

write. This paper provides a view of selected issues relevant to a brain-centered psychology, and at the same time relevant to psychoanalysis. I will pursue the following sequence: (1) A condensed account of the history and definition of the term *consciousness*. (2) A discussion of consciousness in terms of information theory and the notion of *re-representation*. (3) A discussion of the evolution of consciousness and its part in the larger context of human evolution. (4) A brief description of the currently known neurobiology underlying consciousness. (5) An exploration of the relevance of the preceding issues to the psycho-analytic explanations of defense and the effects of psychoanalytic therapy.

History

The idea of consciousness has not always been with us. Among the ancients, there was little idea of consciousness as we think of it. The idea of the divided self in the sense of an internal homunculus or soul controlling a body had not yet developed. The Greeks had many different theories of mind, so it is difficult to generalize, but a popular conception involved a monistic system of *psyche*. Events that we might call unconscious, mythic cultures usually attributed to divine intervention, to a god inducing a dream figure to appear to a human, or to a god inducing irrational behavior. With Plato appears a more modern idea: the appetitive, lustful faculty and the spiritual faculty are subordinated to the rationale faculty; except in sleep, when the lower faculties escape from rational control and may bring dreams of monstrous behavior. Here we find a sense that forbidden wishes may originate from within; yet even Plato gives us no concept of consciousness and unconsciousness or any notion of self-awareness and introspection of the modern sort. The Middle Ages return to a more mythic approach — intrusions by God, influence by the Devil, and the like.

The word *consciousness,* according to Whyte (1960), begins to appear in the seventeenth century in Europe. Descartes took the long-established split between *divine soul* and *mortal body* and secularized it to render a duality between mind and body, which we have inherited. This dualism has become part of our "common sense," with the resulting contradiction that within our apparently unified universe are two kinds of substance. The intuitive sense that there is a "self" or "mind" over against a body is impossible to remove from our daily thinking, where this contradiction does not seem to matter. In scientific thinking we can ignore the split until we come to psychology. Here the problem will not go away.

This dichotomy has also pervaded the thinking of psychoanalytic theorists. From the topographic model, where the mind was simply split into an executive consciousness and a drive-influenced unconsciousness, to the structural model, where the executive ego had its own unconscious element, this dualism returned in different guise. One of the aspects of the Freudian revolution troubling to traditional philosophy was that one had to confront the fact that the unconscious, with its feet in the clay of *res extensa,* could influence the commonly conceived nonmaterial conscious self. Freud saw consciousness as an organ different in type from the rest of the mind, although not necessarily of an immaterial type. He connected consciousness and perception as organs that could receive sensible qualities but did not store memory traces. In the Project (1895) he called this the *omega* system, in later writing he (1900, p. 615) named it the perception-consciousness system, which received sensory information from the outside world but which also had an inner view as the *"sense organ for the perception of psychical qualities."* He related its internal aspect importantly with "word-presentations" as opposed to unconscious "thing-presentations" (see Natsoulas, 1989), a prescient understanding in view of the current emphasis on the importance of language to consciousness.

The fact is that in the early part of this century *consciousness* was given scant attention. For the psychoanalytic movement the development of the theory of the *unconscious* was of more pressing interest. On the other hand, in the world of academic psychology, important work on consciousness was begun in the nineteenth century by major figures, such as Wilhelm Wundt and William James; but it came to a halt with the hegemony of behaviorism. Well into the 1940s, "consciousness" was an unscientific form unworthy of research attention. In the 50s, however, such diverse workers as Broadbent (1958), Lashley (1951), and Chomsky (1957) broke the behaviorist grip, and the study of consciousness is currently of major importance.

Recent advances in neurobiology and information theory have generated new models of mind in the psychoanalytic realm, that have eschewed the Freudian drive theories underlying the earlier dualistic system. All the versions of the Freudian model included id-generated psychic energy, orchestrated by "higher" agencies that control and civilize behavior, rather like the Platonic model. In later versions, as Rosenblatt and Thickstun (1977) have pointed out, an informational nuance crept in under the name "signal anxiety," really a bit of information that warned of danger and the need for inner defensive measures. Rosenblatt and Thickstun, Peterfreund's informational model (1971), and Basch's paper on communication science (1976) have been major influences in the recasting of psychoanalytic theory.

Dualistic models still exist. It is my view that such models severely strain credulity, requiring as they do two kinds of substance, and render any psychological system incomplete. The behaviorists, wrong-headed as they now seem, did perform the service of unifying the split by legislating *consciousness* out of consideration. My endeavor has been to instead use the concepts of information and semiotics to explain why the split is only an apparent one. I have discussed this at length in the previous article: I will briefly review the arguments later in this paper, as I try to include consciousness in an information-processing model.

Definitions of Consciousness

Definitions of consciousness are strongly influenced by one's theory of consciousness and vice versa. Consciousness is one of those complex concepts which draws one into a vicious cycle of definition and redefinition.

Consciousness is often defined by what it is *not,* by various kinds of unconsciousness, the most basic of which is the opposition to stupor, coma, or sleep. At another level we oppose it to vague states of inattention such as trance, fugue, or even distraction. A third is based on awareness. One may be aware enough to be engaged in a familiar activity, such as driving, or cooking, but simultaneously thinking about something else. At a higher level one may be focused fully on a task that is new or difficult without being "self-conscious." Then, usually at the top of the consciousness hierarchy, we place self-consciousness, wherein I am aware of what I am doing and notice the fact that *I* am doing it. I will deal here with the last three of these definitions, all involving awareness and self-awareness; they have a reflective or recursive aspect, which makes them unique, and which I call *re-representation.*

We can define consciousness in terms of *content* or in terms of *process* (Rosenblatt, 1990). The content definition deals with what is or can be included. The Freudian model included the currently conscious, the potentially conscious or preconscious, and the unconscious. The unconscious also must be divided between the dynamic unconscious, the contents which can not be brought to awareness for defensive psychological reasons, and those which involve brain and body functions, such as motor and autonomic regulation. The *content* definition is a *functional* one, which addresses the "what" and the "why" of consciousness. The *process* definition relates to the "how." This would ideally include a neurobiological explanation; but we are not yet ready to do this. My hope here is to use an informational model to lay some of the groundwork for future speculations about this process.

The Role of Consciousness in Information Processing

Essential to our discussion is the idea that the brain is an information processor. Claude Shannon (1949) formulated an information theory that has become standard as the basic theory of communication systems. The theory uses the laws of thermo-dynamics as a model for informational structures and permits a kind of quantification of information by use of the concept of entropy; like any structure, information bears an inverse relation to entropy.

A corollary to this theory, which was important to Shannon, is the fact that messages are structures that tend to degrade, as in the game of Telephone. All effective information systems have methods for preventing this degradation. For instance, in human language there is a great deal of redundancy so that one can often understand a sentence even when one hears only half the words. Electronic messages have several tricks for reducing error, such as retransmission or providing digit counts. Or, in other communication systems, sending the same message by different media can be effective, such as using both blinking-light Morse code and semaphore between ships at sea. One purpose, then, of sending a message more than once—*re-representation*—is for validation.

An example of a system *without* a re-representation device, which is consequently a poor system, is a dial telephone. If I dial a number and no one answers, I have no way of knowing if I dialed the wrong number or if indeed no one is at home. I may do a primitive verification trick and dial again; but a much better system is a phone with a digital read-out that shows just what numbers I dialed.

Validation in Biological Systems

Let us consider biological information systems to see if they contain examples of re-representation. One example might be the way enzymes organize the pairing of nucleic acids in making copies of DNA molecules. At one end of the enzyme the pairings are made, at the other end the sequence is checked for errors and these errors corrected. This also is a re-representation system, a kind of proof reading of the gene.

With cybernetic biological systems, the re-representation is a feedback *sign* to the governing entity. For instance, blood-sugar exists in the blood; it also becomes a sign of its own level by being counted, compared to standard, and thus stimulating an error correction. Or, a flying bee keeps re-representing its distance from the hive and flying toward it to correct for the error (distance).

Representation as Sign

Let us clarify the idea of "representation." In my previous paper, I proposed that the theoretical discipline *semiotics* has something to offer in explaining mind-brain interaction. The theories of Poinsot (1632), Peirce (1897), and Sebeok (1986) converge on the idea that the *sign* is ubiquitous in all living systems. Basically, these theories suggest that all life depends on the ability of one thing to *stand for* another. An amino acid stands for a codon in an RNA molecule, an RNA molecule stands for a DNA molecule, glucose concentration stands for a certain metabolic state, etc. With respect to our current interest, brain events (more or less permanent cellular modifications) *stand for* percepts and by so doing can "store" them in memory. Other sign systems can *stand for* these brain-storage events. Human language is one such system. I mean that words, expressed in appropriate syntax, stand for memory traces stored in the brain. We may say they carry the information for the purpose of communicating it to others. Similarly, gestural language, sign language, writing, music, and others are systems for translating brain events into signs that may be transmitted. My thesis is that consciousness is one such sign system; *it is a sign system in which cellular events in the consciousness system re-represent cellular events that have recorded information in the nonconscious brain.*

Consciousness as After-the-Fact

Let us pursue the idea that consciousness is part of such a re-representation system. Traditionally, consciousness has been placed in the position of initiator of behavior, the decision maker, the center of will. The work of Libet et al. (1983) brought this idea into question by revealing that conscious awareness is after the fact. It is well known that a reflex withdrawal after stepping on a tack will be *followed* by consciousness of that act. But Libet and his colleagues demonstrated that even with an active spontaneous decision a *readiness potential* in the brain can be detected almost a half second before conscious awareness of the decision. So all computations and cogitations would seem to be initiated by the nonconscious brain; some of these are fed into the narrow stream of consciousness.

The conclusions derived from this finding are at first glance distressing. It suggests that we do not make conscious decisions; therefore the whole system of personal responsibility seem to be in jeopardy. But in fact the conclusion is almost self-evident even without the experiment of Libet et al. If we think of the alternative to this conclusion, namely, that consciousness can be a cause of brain activity, then we have to accept that consciousness is uncaused or comes from outside the brain; this assumption forces us into a dualist position.

Also, we have to remember that the Libet et al. experiment represents a microscopic view—the events taking place in 500 milliseconds. When we make any decision that really counts, that has any moral implications or touches on issues of responsibility, the time frame is much longer. A prolonged *dialogue* takes place between brain and its self-monitoring consciousness; it should be possible to hold the entire person responsible for the result.

This leads us to think about some other curious aspects of consciousness. One is that when we speak we can be conscious of the words we speak, we can even explain how we constructed the sentence (in retrospect), but we can not be conscious of our use of syntax as we are speaking the sentence. Another related fact is that most thinking and reasoning is done outside of consciousness. Research psychologists of the Wurtzberg School understood this as far back as 1900. In Karl Marbe's experiment, reported in 1901 (discussed in Jaynes, 1976), subjects were asked to lift two weights, one in each hand, and place the heavier one in a designated spot. It was clear that the instructions were consciously received, the result was consciously perceived, but the act of judgment could not be retrieved by any amount of introspection.

Similarly, the most creative complex decisions are not made consciously. Kekulé's solution to the benzene-ring problem involving his snake dream, and Poincaré's dazzling insight into Fuchsian functions while stepping onto a bus, are examples of this phenomenon.

Evidence suggests then that consciousness *is* a re-representation device. The brain replays some of the results of its activity—later and in a different medium. Subsequently, in replaying a scene from memory, one becomes aware of what one has learned and can then make corrections. This capacity is similar to the computer monitor, which, like the telephone read-out, tells me what I have just typed. If I did not have the monitor I would not be sure of exactly what I had written. I could perhaps print it out, but if that printout had to be my final draft, I would be typing with much anxiety and loss of efficiency and speed.

Let us look at this phenomenon a little more closely. What needs feeding back? If we follow the computer metaphor further, in its function as a word processor we can find primitive analogies to some thinking processes. In terms of feedback, the monitor shows me what I have just typed. I can make corrections in spelling, punctuation, as well as in the meanings of sentences, paragraphs, even the concept of the whole manuscript. In other words I may make changes in the signs themselves (correct spelling errors) or in the meanings I wish to generate (changing the words). The monitor, however, is only a part of the feedback loop; the human operator is the most essential part. The computer can not write a word by itself.

To go one step further with this analogy, the computer screen displays only the results of the machine's computations. We do not see the computations themselves. If a computer were designed so that each computation, presumably done in machine-language symbols, were represented on the screen, and that the process was slowed down sufficiently so one could see it, there would be considerable time and space between each final character on the screen. Consciousness seems to be a series of tips of icebergs, discontinuous frames showing the results of computations. It may be that the mind turns them into a continuous experience, as it does with the frames of a movie film.

Feedback and Re-representation

One of the essential principles of living systems is that all processes are governed by feedback loops, which themselves include sign systems. Genetic processes are governed by the "feedback" principle of survival. Within organisms metabolic processes such as glucose metabolism, hormone regulation, and muscle control all remain in balance because of negative feedback. In fact, for a process to work, the feedback is as important as the process itself. Without proprioceptive mechanisms, muscular action becomes chaotic. Disruption of hormonal feedback systems can lead to death. Manipulation of the feedback system can also be useful in certain circumstances—for instance, in altering the output of endocrine systems, in using hormones to slow the progress of cancers, or in manipulating neurotransmitter levels with psychotropic drugs.

If feedback systems are ubiquitous, then we can enquire whether they have something to do with consciousness. With perception, a multichannel input from the senses is organized and routed to consciousness so that we are aware of our immediate experience. In the loop of input-output the conscious representation is itself a feedback phenomenon, which the brain takes into account in forming the next action or response. Analogously, when we recall an image from memory, again representing it via multiple channels, often with attendant affect, the conscious representation acts as the next input to the brain. It is the sign to the brain representing what the brain has just accomplished.

The interesting thing is that this re-representation is a highly selective linear stream out of the massive parallel-processing brain. Mandler (1988), in discussing consciousness, stresses that the conscious stream is not only narrow and selective but that it is a constant process of construction, of putting together the multiple input trains into an ever-changing complex. The point for our purposes is that the selection is for *items of interest for feedback*. It is a sampling of brain activity, not of the whole of reality. It resembles a diagnostic device like that of taking one's pulse to see if one is doing an adequate workout. Similarly, an airline pilot, instead of looking out the window, may observe his instruments for read-outs concerning the position or altitude of his aircraft. Consciousness does not "look out" through the eyes (and other senses); it represents what the brain has assembled from using all the senses.

The choice of what to represent consciously is a puzzle that tends to generate a circular argument. It is conscious because it is deemed important; it is important because it is conscious. Ultimately we need a mechanism to explain how *importance* is evaluated and how importance leads to consciousness. The currently exciting neural-net models of the mind (Campbell, 1989) suggest possible mechanisms involving alterations in connection strengths of certain complexes to put them over a hypothetical threshold to consciousness. In this kind of model, that event itself — the becoming conscious — alters relationships in proximal or "associated" complexes, which may then be boosted over the threshold.

As to the reasons for the supraliminal importance of conscious contents, we have to assume that certain hierarchies exist. In the realm of *content* there may be ordering in terms of survival value, instinctual importance, interest, fantasy, etc. With respect to structural qualities of perception there are other "attention-getting" criteria, such as intensity of stimulus, novelty, or dissonance. We can only surmise the reasons for choosing a particular conscious item, and we often do just that in psychoanalytic therapy. Lichtenberg (1989) has described motivational systems underlying content hierarchies.

Neurobiology of Consciousness

As I mentioned in attempting an initial definition of consciousness, we think of the term on a hierarchical series of planes. And these planes all have neurological correlates. Kissen (1986) describes the known circuits that are considered responsible for the functioning at the different levels.

The most basic definition of consciousness opposes it to such states of unconsciousness as sleep or coma. From the neurophysiological point of view this basic distinction is quite complex. There are thought to be two primary activating systems in the brain stem, one mediated by norepinephrine, the other by acetylcholine. The *nor-adrenergic* system involves two anatomical structures — the reticular activating system, which drives basic states of arousal, and a subdivision of that system, the locus coeruleus, which organizes states of attention. The other activating complex, the *cholinergic* system, stimulates the brain but does not generate alertness; this is the mechanism that is active in REM sleep.

At a higher level there is a *general awareness system* that mediates an awareness of experience and surroundings, as opposed to focused attention. This involves a system which includes — and apparently links — parts of the hypothalamus, the thalamus, and the basal ganglia (globus pallidus and putamen). Damage to parts of this system can produce such syndromes as akinetic mutism in which there is wakefulness but minimal reaction to any stimulus, or the thalamic syndrome of Dejerine-Roussy, in which emotional reactions to stimuli are grossly exaggerated or diminished. Thus the brain centers at this level are necessary for an integrated but unfocused sense of being in the world, along with a general affective sense of one's condition.

Self-Awareness System

When we come to the most "advanced" capability of consciousness, which we may call a *self-awareness system,* we have to deal both with a high level of consciousness and the idea of *self.* Self is a concept that has filled many volumes; here I will limit my discussion to the minimum and deal with it in relation to our information-processing model.

Several neurological deficit syndromes have led us to clarify and refine what we mean by conscious self-awareness. The best known such deficit is that arising from splitting the brain by cutting the corpus callosum. Such lesions have led to the discovery of specialization of functions in the two hemispheres, the results of which are by now generally known (Gazzaniga, 1978). Of most interest in the present context is the fact that visual information fed to the isolated right hemisphere alone will not enter any kind of self-awareness; the person will deny seeing anything. At the same time experiments have shown that the object is perceived and registered at a certain level. Although the person can not name the objects she can point to them or rate them according to feelings about them, such as whether she likes them or not. In one experiment, the picture of a nude man was presented to the isolated right hemisphere of a woman. She said she saw nothing, but at the same time she blushed and giggled. In such situations we have to assume that the person receives the images up to the level of the right parietal cortex and that there is some sense of *self* in relation to the objects. But the last step into the linguistic brain is missing. Self-awareness seems to require that step.

The work of Mountcastle et al. (1975) and Mesalum and Geschwind (1978) has elaborated on other complex connections underlying the focused awareness of the self-awareness system. Experimental lesions of the posterior-inferior area of the parietal lobe in monkeys lead to an inability to focus attention on the contralateral side. In humans cerebrovascular damage to the same area leads to the unilateral neglect syndrome in which the patients cannot even conceive of anything of interest on that side of the body, or in the world, that would be perceived in that visual field.

The "self system" seems to involve integration of many inputs at the posterior inferior parietal lobe — by routes from the other association cortical areas, from the limbic system and brain stem. This is a major integration point of somatosensory and affective input; the sense of one's body appears to be central to the sense of self. The important conclusion from the research into the neglect syndromes is that the self system is as essential to focused awareness as are the intact perceptual systems and lower awareness systems. To focus on objects they must be seen as *significant* to a self. Mesalum and Geschwind (1978) have concluded that the self system is somewhat more localized in the right brain than in the left. According to this definition of self, selfhood exists in all mammals; if it is obliterated a neglect syndrome will follow. So the sense of self developed early in evolution, long before consciousness in human terms. This self exists on conscious and unconscious levels, integrating proprioception, sensation, and affect into a central entity, which becomes a center of *significance* and mediates the importance of all experiences.

As we review the neurobiology, it becomes possible to imagine that the mechanism of consciousness involves certain brain centers. It may be that there are multiple centers, perhaps even a diffuse distribution. But in the discussion to come let us accept that conscious events are brain events, involving discrete areas of the brain, without worrying about exactly where they are. To conceive of consciousness as a sign system allows us to proceed without exact anatomical knowledge.

The idea of a sign system also helps with the distinction between *consciousness* and *self-consciousness*. Consciousness has to do with re-representing percepts and memories selected from the brain's vast storehouse. Self-consciousness has to do with re-representing those signs which refer to the individual as an object and as a center of significance.

The idea of a "consciousness center" that re-represents brain events in a feedback loop by way of selective integration, is, I believe, useful heuristically, although it does not land us in conceptual utopia. Vexing complexities arise immediately. One is the *when* of consciousness. If consciousness is a reflective loop or feedback system, does consciousness of a percept or a memory occur at the instant it reaches the hypothesized consciousness center or when it returns to wherever in the brain it originated? Another question is what chooses the contents of consciousness, the whole brain, or the consciousness center? If we say the whole brain, we seem to be begging the question, since the brain, in order to know what to flash into consciousness, must be aware of it already. Contrariwise, the consciousness center is in the same position. These are problems that have long plagued the notion of the "sapient ego." The answers, emerging from neural-network theory and related philosophical writings, suggest that the brain functions in a modular fashion, with "multiple partial homunculi" (see Dennett, 1978). The general idea is that if we have specialized homunculi who are not simply miniature versions of the self but are specialized members of a team that "knows" more than its individual members, we do not have the problems of infinite regress, and, in this case, who chooses what becomes conscious. The parallel-processing brain sends certain items to the consciousness system, which is itself a brain area specialized for re-representation (see Anscombe, 1986; Gazzaniga, 1985).

Consciousness in Evolution

If we look back through the history of evolution we may be struck by a phenomenon that has interesting similarities to consciousness, namely, the REM state in mammals. In his book *Brain and Psyche,* Winson (1985) describes an evolutionary leap in brain ability. The most primitive order of mammals, the monotremes, are mammals in that they have a four-chambered heart, are warm blooded, and suckle their young; but they still have a reptilian vestige: they hatch their young from eggs. Another striking thing about them is that their brains are larger and more convoluted for their body size than those of much more intelligent animals further up the evolutionary tree. Noting that the monotremes, unlike later-evolving marsupials and mammals, do not have REM sleep, Winson postulates that REM sleep is an efficient organizer of brain activity, a mode of "off-line processing" of information; this efficiency may allow for a relatively smaller brain. He theorizes that in sleep, mammals re-represent complex behaviors, possibly as a form of rehearsal or of maintaining skills, perhaps the forerunner of the "fantasy rehearsal" currently popular among athletes.

Normally the motor system is paralyzed during REM states; if this paralysis can be surgically prevented, the animal will move its body as if it were running and leaping. The point for our purpose is that the REM dream may be the first evolutionary instance of *cerebral re-representation*. This is not the same as recognition memory such as exists in a reptile, enabling it to recognize stimuli, learn tasks, and follow elaborate operant procedures. In such an animal an engram in the brain is activated by the recognized stimulus; but we presume that the reptile cannot call up a fantasy of that stimulus and run through an action sequence in fantasy. With the behavioral learning of such creatures, the memory consists of a recognition of a stimulus followed by the responsive behavior. With the mammal's REM state, on the other hand, the episode can be expressed in representative form, without the behavior.

Let us speculate that the ability to re-represent in image or trial-action form is an achievement of lower mammals and that it may be analogous to what we call consciousness. The next evolutionary stage would be the ability to do such imaging and trial actions in a *waking* state. Apparently other primates can perform complicated tasks and can plan complex strategies. For the chimpanzee who fetches a box to climb upon in order to reach the banana, it is likely that some kind of cerebral trial action took place in forming the action strategy.

Thus there is a phylogenetic line of self-consciousness-like phenomena. The most primitive organisms have biochemical feedback systems, such as those which represent glucose levels, hormone levels, and response to antigens. With more overtly informational processes feedback is still required. In an animal communicating through facial or bodily gesture, in the fashion Darwin described, there must be internal feedback systems to let the animal "know" it is generating the right gesture, as well as the external feedback from those who receive the signals. Such feedback systems are unconscious, for the most part. With language we must hear ourselves speak to know what we said; this too may be more or less conscious. Self-consciousness is like all the other forms of feedback in that it is a re-representation with certain additions including an owning, a self-reference, or self-responsibility.

Language and Consciousness

How necessary is language to consciousness? Given the split-brain experiments that show patients having perceptions by way of the right hemisphere, but showing no evidence of consciousness of them, language seems to be necessary (Gazzaniga, 1978; LeDoux, 1985). However, a commonsense objection arises. One can imagine visual scenes without the need for language; in fact a scene image may be so elaborate as not to be fully describable in words.

Gazzaniga (1985) describes experiments that show that in even those split-brain patients who have some right-brain language ability, it is the left brain that is better able to call up scenes in imagination. If this capacity is a left-hemisphere phenomenon it may be related to language, and the following hypothesis emerges: that the imaging capacity was an evolutionary precursor to language. In other words, the functions necessary to language, such as re-representation in a linear sequence, analogous to a spatially imagined sequence and involving a linearly ordained logic, required an advanced imaging capacity. This becomes even clearer when applied to written language, which makes full use of the power to imagine visual symbols in a sequenced, syntactic order. Thus it seems that consciousness, in order to be fully rigged, requires both the imaging and the linguistic capacity.

We are now in a position to guess at an evolutionary series leading to consciousness as we know it. First were biochemical feedback systems, then, with motile organisms, proprioceptive feedback systems. With early mammals we have a primitive visualization system operating in sleep, the REM dream state. Next comes a capacity to visualize scenes and form strategies, from the behavioral map that allows a dog to find its buried bone to a complex strategy that helps a chimpanzee to reach a banana.

The awake visualization becomes subspecialized into the language functions in the dominant hemisphere, this final set of systems making up the basis of our consciousness system.

Co-evolution With Other Functions

Having focused so far on the re-representation aspect of consciousness, we must now confront the fact that consciousness as re-representation exists within a group of functions that probably co-evolved, each one necessary for the functions of the others. We have related consciousness to the self, to language, and to society. It is probable that in order to have any one of these we must have them all. As they developed in tandem each must have synergized with the others, leading to a system inextricably bound with the development of the individual in society.

We may be able to discern at least four of these functions.

1. The *imaging* function has the longest genealogy. It evolved as an advanced form of re-representation, folllowing a history that includes biochemical feedback, proprioception, the REM dream, and advanced imaging systems.

2. Another is the *self* function, also with primitive origins but becoming more complex as the organism has advanced and the imaging function allowed for a *self image*.

3. A third is the empathic function, most likely a late developer. Humphrey (1984) theorizes that consciousness of one's own feelings is helpful in judging or predicting the feelings of others. This ability might give one a selective advantage in the community.

4. Fourth, and presumably the latest to emerge, is the language function.

The possibilities for synergy among these evolving capacities are striking. The imaging function allows for an enhanced sense of self, for the sequencing aspects of language, and for the empathic placement of others in predictive simulations of behavior. The self function orients oneself in one's fantasies, allows for empathy, adds to the development of syntax, the subject-object basis of grammar. Empathy allows for images of others to resemble those of self, enriches the sense of self as it is differentiated from others, helps predict behavior, and may help uncover deceit especially that conveyed by language. Language, in addition to its dramatic advancement in communication providing the basis for advanced civilization, allows for the experience of consciousness (as evidenced by the split-brain experiments), stabilizes the self in a linguistic system, and allows for much more elaborate empathic communication.

Defenses: Interference With Re-Representation

As an evolved capacity, we have concluded that re-representation is apparently very useful. The question must arise, why do we so frequently interfere with it?

It is not immediately obvious why we should need to repress memories. It may be, however, that the brain, in order to work efficiently, must protect itself from excessive affect. If we can truly re-represent without motor action, as in fantasy rehearsal, it seems that we should be able to do it without overwhelming affect. However, the two types of separation represent very different processing problems.

In evolution, the perceptual system has a long history of division from the motor system. In reflex behavior we say there is no separation; action follows percept with no intermediate processing and no choice. As time has gone on, however, higher organisms have been able to perceive and then make use of intentional systems, having some choice in how they will make use of the perceptions.

It is as though we have been able to achieve trial action but are only partially successful at "trial affect." The brain is wired so that every percept comes with some affective loading. Unlike action, which is an output phenomenon and therefore can more easily be separated from input, affect is part of the input. The information-processing task is that much more difficult because the affect must be isolated *after* it has arrived packaged with the percept. The defenses are used in various ways to make this disconnection. Why? One reason may be that internal information processing is difficult when there is no separation. Certain affects are themselves aversive. The memory of a trauma brings with it a flood of painful affect which, in the absence of defenses, would be overwhelming, precluding any further thought or planning. The repertoire of defenses seems to be a graded set of methods for keeping consciousness as free as possible from overwhelming emotion and frenzied reaction.

The individual's motivation to defend against negative affect is operantly conditioned — avoidance results in less pain. This process of affect avoidance can be maladaptive, and can lead to symptoms. However, in terms of evolutionary adaptation and natural selection, the defenses may be useful in the overall scheme of information processing. *Defenses are analogous to the muscle paralysis of REM sleep;* they have evolved to allow for conscious function at least partially independent of affect.

The idea of consciousness as a feedback loop helps to clarify this conceptualization of defenses. For one thing, the fact that it is a loop means there is an efferent and an afferent limb. We could predict that there may be blockages possible with respect to each limb. Repression and the other defenses devoted to keeping ideas and their attendant affect out of awareness may represent a functional interference with the transmissions *to* the consciousness center. When certain mental content has reached the consciousness center and is on its way back, it may reach a state of awareness but be disavowed or split. This leads to interesting questions. If we could time defenses in an experiment similar to that of Libet et al., would we find disavowal tending to occur some milliseconds *later* than repression?

The Function of Psychotherapy

Being able to separate and repress affect and allow for unencumbered thinking in a crisis has evolutionary advantage. In fact it is no doubt beneficial to a social organism to repress and otherwise defend against various impulses that would be disruptive—Freud's (1930) model of civilization. In many people these defenses may be excessive; they can become chronic maladaptive aspects of personality, leading to problems in information processing throughout life. The results of childhood conflicts may be kept out of consciousness; they become set in place with no opportunity for revision. Once repressed ideas can enter the conscious stream, they can enter into a process by which conscious ideas are fed back into the preconscious mind and can be evaluated in relation to the person's adult understanding of the world.

In this light psychoanalytic therapy is an advance in social evolution that helps some individuals to succeed better than they otherwise would. A common rationale for therapy is that we use the defensive systems when they are needed in childhood, but that when their usefulness has passed we may be crippled by them. Therapy is then required to help undo the obsolete defenses. The development of psychotherapy could be seen as a social development, similar to prolonged child care, the use of clothing, and certain urban amenities. It exists in the tradition of most of the world's religions, although most religious techniques tend to reenforce defenses rather than try to undo them. In societies in which life is much harsher and more dangerous than in the twentieth-century West, the defenses remain necessary more than they do for us; or it may be that *different* defenses are required. In societies with surplus capital, leisure, and time for reflection and creativity, the undoing of certain defenses may release consciousness and expand it, to adaptive advantage.

Psychotherapy is an analog of consciousness, a kind of *re-re-representation,* which involves a projection of dynamic unconscious structures onto the analyst, who reflects them back. Of course other things happen in therapy by way of internalization, new behavioral learning, and cognitive restructuring (Olds, 1981). These things happen because the analyst takes the next step beyond consciousness.

The analyst or therapist, in taking a neutral, nonjudgmental role, is performing a function similar to the feedback function of consciousness itself, allowing for information to be brought into the discourse with the affective aspect temporarily reduced. In fact this is never entirely possible. But in a successful therapy such a condition may be approached as the patient begins to trust the analyst's noncritical stance and realizes the therapist is benignly neutral.

Discussion

To summarize, in the process of evolution we have developed information systems devoted to processing and error correction, which have certain things in common. These systems allow for an escape from the constriction of behavioral learning and performance, in which behavior is learned always on the hoof, in the process of action. We may surmise that the *dream* is the earliest example of this "off-line" processing, where representations can be brought into some state of awareness, but with the capacity for enactment cut off.

Then we have consciousness, where a similar process can occur in the waking state. Here representations can be reviewed with the motor system disengaged at a higher level—the muscles are ready for action, but controlled by conscious (or unconscious) decision processes; in addition the affect system can be more or less decoupled by use of the defense mechanisms. This allows for self-consciousness, conscious planning, fantasy rehearsal, daydreaming, and all the many products of imagination. The third great system in this series, which also makes use of behavioral decoupling, is psychotherapy, the method of repair. In motorcycle repair, it is useful in making a diagnosis, and in fine tuning, to run the engine in *neutral,* uncoupled from the wheels. In therapy we do something similar. Psychoanalysis is the most obvious in this regard. We ask the patient to lie on a couch, and put the muscles at ease, consistent with a waking state. Then the analysand is to disengage from normal practical goals and allow the mind to wander and follow its "free associations." In fact even the normal "observing ego" is told to relax, as well as the observing superego. These functions are handed over to the analyst, with the purpose of allowing free association unfettered by even self-observation. This technique—while never fully realized in that there is never perfect trust or complete relaxation—allows for the revelation of the unconscious structures or inner constraints that guide thought and action. Analysis is a kind of *metaself-consciousness,* with the analyst playing a temporary role of the *metaself.*

Consequently, consciousness itself is a kind of model for psychotherapy, particularly the psychoanalytic varieties of therapy. If consciousness is the repair mechanism for the behavioral repertoire, therapy is the repair mechanism for consciousness.

If we accept this view, its implications have importance: We do not need the homuncular aspects of ego and self psychology. We do not need to postulate an ego or a censor or a superego, except as *virtual* entities representing certain clusters of functions. We do not need a sapient central organizer. It means that psychotherapy is a treatment for the brain. Consciousness reveals the effects of our work. By undoing defenses we alter some of the information that the preconscious brain has to deal with; this may result in the brain's allowing more and different material into consciousness. This change allows a wider range of phenomena to enter the feedback stream and hence may lead to a greater freedom of action.

In a *brain-centered* model we see the organizing entity as entirely preconscious. The organizing entity is the brain itself. The brain seems to be a conglomerate of many semi-independent modular units, which organize themselves in a way as yet not understood. Consciousness is one of the brain's many output phenomena. Other such phenomena include motor actions, visceral responses, words, and dreams. The parallel-processing brain does thousands of tasks simultaneously. Each of its output processes is characterized by a reduction to a linear series. With motor activity there is a coordinate stream of action in a chained sequence. Proprioception gives us instant feedback regarding our progress along the chain. Conscious thought is the selective activation of a stream of current percepts, recalled images, or thoughts. This stream represents what has just happened in the brain, and acts like a video monitor. Each image or thought that becomes conscious acts as another input to the brain. Consciousness is a kind of proprioceptor of the thought process, a way of the brain's informing itself what it has just accomplished.

What are the implications for our theory of therapy? One result is an integration of our techniques. The verbal, behavioral, and pharmacologic therapies can be seen as altering different kinds of input to the preconscious process (Olds, 1981). I am not implying that there should be changes in *psychoanalytic* technique. In fact this model gives considerable theoretical support to the analytic method, especially its empathic, reflective aspects. However, this theory does imply that psychoanalysis is one of several modes of modifying a person's pattern of information processing and that a combination of psychoanalysis with other treatments may in some cases be desirable.

In semiotic parlance conscious thoughts, words, and actions are *signs* of brain activity; they represent a sign system by which the brain talks to itself. They are also part of a sign system in the interpersonal world, in the culture in which one lives. In this conceptualization, neurotransmitters, pharmacologic agents, autonomic responses, words, and gestures are all aspects of sign systems; they can all act as inputs to the brain in its preconscious activity.

If we expand on the feedback-loop notion this point becomes more emphatically clear. If consciousness involves a brain center that integrates information from several brain systems — cognitive, behavioral, and affective — then a therapy that promotes a freer flow and widening scope of unconscious brain contents into consciousness will alter all these systems. Schwartz (1987) points out how the altered expectations of the analytic situation — the neutrality, abstinence, "containing" environment — alter a host of associational connections, connections between ideas, habits, behavioral expectations, and affects. Analysts tend to view what they do as mostly verbal, but the analytic experience is a global change in experience, with immediate effect on mood, reactivity, and opportunities for response. These lead to biochemical changes in the affect system, to habit changes in the behavioral system, and cognitive changes in the systems of internalization and verbal learning. Some of these changes open up the afferent limb to the center of consciousness, some the efferent limb. In neural-network theory any chemical alteration in the neurotransmitters involving affect or in those mediating other kinds of learning must modify the stream into and out of consciousness. Just lying on the couch will make a minuscule change in affect, which will modify the ensuing chain of associations. Similarly any action of the analyst, including any interpretation, since it is carried out in an otherwise low-noise atmosphere, will also alter the stream. As this process leads to a deepening transference process, its course changes even more, and new defended-against and/or forgotten material catches the "mind's eye." The stream of consciousness becomes diverted like light waves around a planet, bending toward the transference.

REFERENCES

Anscombe, R. (1986). The ego and the will. *Psychoanal. & Contemp. Thought,* 9:437–463.
Basch, M. F. (1976). Psychoanalysis and communication science. *Annual Psychoanal.,* 4:385–421. New York: Int. Univ. Press.
Broadbent, D. E. (1958). *Perception and Communication.* London: Pergamon.
Campbell, J. (1989). *The Improbable Machine.* New York: Simon & Schuster.
Chomsky, N. (1957). *Syntactic Structures.* The Hague: Mouton.
Dennett, D. (1978). *Brainstorms.* Cambridge: MIT Press.
Freud, S. (1895). Project for a scientific psychology. *S.E.,* 1.
_____ (1900). The interpretation of dreams. *S.E.,* 5.
_____ (1930). Civilization and its discontents. *S.E.,* 21.
Gazzaniga, M. S. (1978). *The Integrated Mind.* New York: Plenum.
_____ (1985). *The Social Brain: Discovering the Networks of the Mind.* New York: Basic Books.
Humphrey, N. (1984). *Consciousness Regained.* New York: Oxford.
Jaynes, J. (1976). *The Origin of Consciousness in the Breakdown of the Bicameral Mind.* Boston: Houghton Mifflin.
Kissen, B. (1986). *Conscious and Unconscious Programs in the Brain.* New York: Plenum.
Lashley, K. S. (1951). The problem of serial order in behavior. In *Cerebral Mechanisms in Behavior,* ed. L. A. Jeffries. New York: Wiley.

LeDoux, J. E. (1985). Brain, Mind and Language. In *Brain and Mind,* ed. D. A. Oakley. London/New York: Methuen.

Libet, B., Gleason, C. A., Wright, E. W., & Pearl, D. K. (1983). Time of conscious intention to act in relation to onset of cerebral activity (readiness potential): The unconscious initiation of a freely voluntary act. *Brain,* 106:623–642.

Lichtenberg, J. D. (1989). *Psychoanalysis and Motivation.* Hillsdale, NJ: The Analytic Press.

Mandler, G. (1988). Problems and directions in the study of consciousness. In *Psychodynamics and Cognition,* ed. M. Horowitz. Chicago: Univ. Chicago Press.

Mesalum, M. M. & Geschwind, N. (1978). On the possible role of neocortex and its limbic connections in the process of attention in schizophrenia: Clinical cases of inattention in man and experimental anatomy in monkey. *J. Psychiat. Res.,* 14:249–259.

Mountcastle, V. M., Lynch, J. C., & Georgopoulos, A. (1975). Posterior pariental association cortex of the monkey: Command function for operations within interpersonal space. *J. Neurophysiol.,* 38:871–908.

Natsoulas, T. (1989). Freud and Consciousness: IV. *Psychoanal. & Contemp. Thought,* 12:619–662.

Olds, D. (1981). The behavioral schema: An integration of modes of learning. *Psychiat.,* 50:112–125.

———— (1990). Brain Centered Psychology: A semiotic approach. *Psychoanal. & Contemp. Thought,* 13:331–363.

Peirce, C. S. (1897). Logic as semiotic: The theory of signs. In *Semiotics: An Introductory Anthology,* ed. R. E. Innis (1985). Bloomington: Indiana Univ. Press, pp. 1–23.

Peterfreund, E. (1971). *Information, Systems, and Psychoanalysis. Psychol. Issues,* Monogr. 25/26. New York: Int. Univ. Press.

Poinsot, J. (1632). *Tractatus de Signis: The Semiotic of John Poinsot,* tr. J. Deely. Berkeley: Univ. California Press.

Rosenblatt, A. (1990). Discussion of this paper at the Annual Meeting of the American Psychoanalytic Association, May, 1990.

———— & Thickstun, J. T. (1977). *Modern Psychoanalytic Concepts in a General Psychology. Psychol. Issues Monogr 11. New York: Int. Univ. Press.*

Schwartz, A. (1987). Drives, affects, behavior—and learning: Approaches to a psychobiology of emotion and to an integration of psychoanalytic and neurobiologic thought. *J. Amer. Psychoanal. Assn.,* 35:467–506.

Sebeok, T. A. (1986). The doctrine of signs. In *Frontiers in Semiotics,* ed. J. Deely, B. Williams, & F. E. Kruse (1986). Bloomington: Indiana Univ. Press, pp. 35–42.

Shannon, C. (1949). *The Mathematical Theory of Communication.* Chicago: Univ. Illinois Press.

Whyte, L. L. (1960). *The Unconscious Before Freud.* New York: Basic Books.

Winson, J. (1985). *Brain and Psyche: The Biology of the Unconscious.* New York: Doubleday.

2211 Broadway, 1H
New York, N.Y. 10024

Not Art But Science: Applications of Neurobiology, Experimental Psychology, and Ethology to Psychoanalytic Technique. I: Neuroscientifically Guided Approaches To Interpretive "What's" and "When's"

ANDREW SCHWARTZ, M.D.

SOME YEARS AGO NOW, A COLLEAGUE observed during a departmental meeting that one of the skills we analysts teach, the interpretive technique essential to psychoanalysis and derivative psychotherapies, is primarily an "art." And this opinion may not be an uncommon one: the catalogue of one long-established institute speaks, for instance, of "a gift for psychological understanding" that "cannot be taught." In his thorough and now classic *Psychoanalytic Technique and Psychic Conflict,* Brenner (1976) enumerates the several ways in which accomplished clinicians describe their selecting the foci of and moments for intervention, and Brenner implies that there may exist neither one single correct

Dr. Schwartz is Faculty, Department of Psychiatry, Office of Training and Standards, D.C. Commission on Mental Health Services, St. Elizabeths Hospital, Washington, D.C.

Material herein was presented as a precirculated paper at the Fall Meeting of the American Psychoanalytic Association, December 17, 1988.

method nor perhaps an underlying shared principle common to the seemingly varied procedures he details.

Prologue—and First Premises

The present approach will offer somewhat differing views. It will propose instead that concepts and findings from neurobiology, experimental psychology, and ethology can help define and identify the raw data of psychoanalysis and may thus make deciding on focus and timing of analytic interventions a truly scientific enterprise. This paper and a companion (Schwartz, in prep.) suggest that countertransference phenomena similarly become more understandable when examined with the aid of current ideas and observations from the neurosciences and related disciplines.

Without question, a helpful locus of departure in any discussion of *principles* of technique would be an accepted definition of optimal *practical* standards, but one might well stand on safe terrain in concluding that no such consensus has yet emerged from the psychoanalytic community (cf. Gray, 1982, 1986; Abend, 1986). Pianists, one might note, have never agreed how best to play their instrument (Gerig, 1974), and the keyboard has been around somewhat longer than the analytic couch—though one should also observe that anatomy and physiology have in recent decades made increasing contributions to the understanding and teaching of musical mechanics. At any rate, Brenner (1976, p. 125) offers helpful first guidelines to goals of intervention when, in discussing timing, he states, "Correct timing means only not long before a patient is ready nor very long after he was first ready."

Significantly, Brenner emphasizes the value of addressing issues as early as feasible, and he thus implies the need to find reliable indices of a patient's nascent openness to interpretation. Furthermore, while Brenner does not explicitly describe what these indicators might be, his emotionally vivid and evocative analytic style (Brenner, 1987) suggests an idea central to this essay's proposals, agreeable perhaps to a considerable spectrum of psychoanalysts, and consonant with landmark and still influential writings of half a century ago (Strachey, 1934; Fenichel, 1941)— that *acknowledgeable affect constitutes a key component of—and may even guide the timing of—useful interpretive comments.*

Writing almost 40 years before Brenner, Fenichel (1941, pp. 45–46) argued that optimal technique comprised interventions that touched "the surface" of the patient's "mind," that in other words conveyed an observation the analysand might have made on his own had he thought to look, and that in sum highlighted data which the listener could readily acknowledge as true about himself. Although Fenichel used the metaphors of metapsychology and thus spoke of the need to address issues "economically" most highly "cathected" or charged with one variety or other of "psychic energy," he clearly, as Gill (1982) has emphasized, was advising that the analyst take up those topics of greatest *emotional* salience: "we must operate at that point where the affect is actually situated at the moment: it must be added that the patient does not know this point and we must first *seek out* the places where the affect is situated."

More particularly, considering "too deep" interventions "naming . . . unconscious processes which the patient cannot feel within himself" (Fenichel, 1941, p. 45), Fenichel (1941, p. 46) observed, "*By no means* is such an 'interpretation' an interpretation in the true analytic sense, which is a real confrontation of the experiencing ego with something it had previously warded off." Today we might wish just to append that Fenichel's reservations could of course pertain equally not only to "too deep" "id"-directed comments but also to mechanistic "defense interpretations" that neglect what the analysand *can* "feel within himself" and that instead impose an abstract theory on the patient and his productions (cf. Brenner, 1976, pp. 21, 64). Furthermore, one should add, a valuable analytic remark may focus on emotional issues more accurately described as "underappreciated" than as "warded off."

At any rate, because Fenichel in effect recommended a focus on clinically evident and *acknowledgeable* affects, he implicitly posed the interesting question to which this essay will shortly turn: *how can we know—scientifically—which feelings are "closest to the surface" and consequently,* in different but rather familiar terms, *"conscious" or "preconscious."*

Before tackling this issue, however, one might note a possibly useful obverse to Fenichel's advice, which one can formulate as follows: *any analytic intervention (1) has an at least implicit emotional syntax and logic; (2) thus assumes,* in clinical terms, *that the patient is experiencing one or more feelings which, in turn, spur him to action and/or reflect events in his inner or external worlds; and (3) will when offered prove acknowledgeable and useful if and only if the affects named or implied in the interpretation are themselves conscious or preconscious and therefore "at the surface."*

With the foregoing ideas in view, then, an analyst can, when considering a given comment, ask himself two questions germane to a *predictive* assessment of that interpretation's current value: "What feelings do my words assume to be at work in my patient?" and "Is there any evidence that my patient is actually feeling those feelings right now?" However, since these queries patently are worthwhile only if one has reliable methods for assessing, almost quantitating, the emotional state of another human being, we reach again that core dilemma: How does one do an "in-the-hour-lab-test" for affect?

Affects "Biologized": Emotions as Neural Appraisals, Motivators, and "Hard Data" Generators

Before even attempting to suggest how it might be possible to identify clinically and perhaps begin to quantify affect in the analytic situation, we must confront a more fundamental question—that of the very nature of emotion—and at this juncture, as observed elsewhere (Schwartz, 1987, 1988), one can usefully consider first the classic senses. Most broadly, the neural systems of vision, audition, taste, smell, and somatic sensation transduce the physical and chemical stimuli of external world and "internal milieu" into electrophysiologic signals—receptor and action potentials—which convey centrally survival-essential information that, by turns, subcortical relays and cortex receive and process in quite organized fashion (Castellucci, 1985; Kandel, 1985b; Kelly, 1985; Martin, 1985).

This transduction and relay process is not random but is instead rather stereotyped. For instance, as Hubel discovered in his landmark experiments with Wiesel (Hubel, 1982; Kandel, 1985a), neurons within the primary visual or striate cortex possess, as do cells of the retina and lateral geniculate, limited and definable "receptive fields"—sections of the total visible area to which they are responsive—and additionally react preferentially to bars of light of specific spatial orientation, direction of movement, and length.

Comparably, in detecting, processing, and registering waveforms and pressure changes in the ambient air, the auditory system and its components demonstrate again that given sensory circuits respond, in somewhat "stereotypic" fashion, only to those afferent data to which these neural elements are sensitive. More explicitly, because of biophysical characteristics of the cochlea's basilar membrane and because of important cell-to-cell differences in ion conductance which endow organ of Corti receptor neurons with increased responsiveness to specific frequencies of vibration, the inner ear can subject sound to Fourier analysis and decompose complex wave patterns into their constituent constant-wavelength sine-waves (Kelly, 1985). Furthermore, cells within the cochlear nuclei, medial geniculate, superior colliculus (Allon & Wollberg, 1978), and other relays exhibit definable tonal "tastes," thus show greatest electrophysiologic reactivity to relatively narrow frequency band-widths, and have, additionally, histologic organization that reflects these acoustic predilections.

Significantly, however, sensory systems not only handle afferent information in stereotyped ways but also tend to trigger relatively invariant sequences of motor activity that include both reflexes of many degrees of complexity and more elaborate behavioral routines known as "fixed action patterns" (Kupferman, 1985c). Examples of these genetically "blue-printed" and fundamentally "hard-wired" motor-programs are legion—for instance, coughing, sneezing, the coordinated precursors of walking observable even in neonates, the startle response to sudden changes in ambient sound, vestibularly-activated righting reactions—but the features to stress here are that these often elaborate yet preprogrammed movements occur following specific kinds of sensation or stimuli and generally have some inferable survival value (cf. Darwin, 1872). While perhaps at first produced by mutational accident, these stereotypies and their underlying "prewired" circuitry have, one may conclude, remained with us as elements of our heritable neuronal machinery because they help us meet and adapt to expectable and enduring biologic and environmental challenges. A newborn infant would encounter serious difficulties if he needed extensive tutoring to learn the intricacies of swallowing.

Given the frequently stereotyped nature of motor response to a particular kind of sensory input, the careful observer can at times derive interesting inferences: if one hears and sees a young child yell and pull a hand away from a stove, one confidently concludes, because of what one knows of patterned reactions to specific sensations, that the burner evoked pain. Obviously, if one takes that deductive step, one has been able *to use audible and visible behavior to gain some understanding of the subjective experience of another human being,* and this latter principle represents one more clearly clinically relevant proposition to which this paper will shortly return.

Still another feature of certain of the classic senses has pertinence to a study of affects and in fact may serve as a convenient conceptual bridge to the examination of emotion to follow. Specifically, while the pathways of vision, hearing, touch, and proprioception produce phenomenologic data of a fundamentally representational and value-neutral nature—these channels can inform us, for instance, if our son did in fact leave his new freestyle bicycle out in the rain, whether the ringing phone does indeed herald the evening's fourth call from our teenage daughter's best friend, or how far our thumb might be from the distracting itch in the small of our back—the circuits subserving pain, temperature, taste, and smell, on the other hand, sample the physical and chemical makeup of our internal and external milieux and offer *appraisals* of our circumstances, which we register as sweetness or salt, caress or slap, perfume or rot. Basically, then, these neurobiologic sensors assess the "quality" (cf. Freud, 1895)—for example, the gustatory or olfactory allure—of our worlds; thus generate survival-essential information; and therefore fit a biologic definition that we may apply easily also to feelings: they *constitute neural processes which allow us to recognize, gauge, and meet the dangers and desirabilities that confront us* (Schwartz, 1990a).

Affects, then, appear to be highly evolved and sensation-like signals with an *appraisal* function (Arnold, 1970; Rosenblatt & Thickstun, 1977a, 1977b; Rosenblatt, 1985), but these phenomena have biologic roots not in the anatomy of the classic senses but rather in structures of the limbic system and related elements of the hypothalamus, upper midbrain, and frontal and temporal cortices (MacLean, 1952, 1967, 1969, 1977, 1978; Kandel, 1985a; Kupferman, 1985a,b). Although to date attempts to define the neuroanatomic substrates of individual feelings have generated only preliminary findings, substantial uncertainty, and not inconsiderable controversy, current data do indicate, for example, that septal, lateral hypothalamic, nucleus accumbens, and ventral tegmental areas house circuitry supporting "pleasure" or "reward" (Olds, 1976; Rolls, 1976; Snyder, 1980; Stellar & Stellar, 1985; Wise & Bozarth, 1985), and the amygdala and medial hypothalamic regions seem comparably implicated in "unpleasure" and defensive affectomotor behaviors (Adamec, 1975; Olds, 1976).

Furthermore, as detailed in other contexts (Schwartz, 1987, 1988), because of productive research into neurophysiologic reward mechanisms, into distribution within limbic-system circuitry of dopamine-containing and opioid-peptidergic neurons (Snyder, 1980), and into the neural bases of "subjectively experienced drug euphoria" (Wise & Bozarth, 1985), one can expand the "appraisal" view of pleasurable affects and suggest that these signals serve additionally as powerfully influential goals of active behaviors. Laboratory investigations of intracranial self-stimulation (Olds, 1976; Rolls, 1976; Stellar & Stellar, 1985) and of drug self-administration (Wise & Bozarth, 1985) mesh both with this inference and with a broader hypothesis proposed elsewhere but relevant also to the present discussion: *appetitively sought hedonic emotions/sensations and aversive dysphoric feelings function respectively as the ultimate "brain-synthesized" positive and negative reinforcers of emotional learning* (Schwartz, 1987, 1988, 1990a).

Obviously, however, affective phenomena extend well beyond subjective experience, and, as Darwin (1872) and other investigators (e.g., MacLean, 1952, 1967, 1969, 1978; Tomkins, 1970; Ekman & Friesen, 1975; Ekman et al., 1983; Ekman, 1984) have observed, each affect, like the sensations earlier discussed, tends to trigger complex motor and hypothalamic-autonomic stereotypies which (1) represent genetically "programmed" and relatively "hard-wired" fixed action patterns originally of evolutionary importance and adaptive value; (2) which via proprioceptive and enteroceptive "feedback loop" pathways generate sensations that in turn promote our learning to identify, distinguish, and label our own feelings; and (3) which finally, through, for example, facial expression and tone of voice, transmit the eminently physical data that, as suggested above, underlie empathy and emotional communication (Schwartz, 1987, 1988, 1990a).

While the foregoing ideas certainly hold for emotions in all ranges of the pleasure-pain continuum, these concepts more particularly allow a refined view of the "unpleasures" that helps put their biologic significance in cleanly-focused perspective: *anxiety/fear, sadness/depression, embarrassment/shame, disgust, and anger/contempt all constitute intricate psychobiological alarm reactions which, by instituting specific motor, cardiovascular, respiratory, endocrinologic, exocrinologic, and other sym-*

pathetic/parasympathetic responses, ready the organism for emergency and crisis, "fight" or "flight" (Schwartz, 1988, 1990a).

Thus, just as do some sensory processes, so does each affect also have its characteristic and specific physiologic expression in which motor and hypothalmic-autonomic stereotypies combine to "write" a recognizable "signature" detectable in facial configuration, muscle tonus, patterned movement, posture, nonverbal features of speech, and vocal timbre. An angry person, for instance, creates a distinctive impression not easily mistaken: clenched jaw, curled upper lip perhaps, bared teeth, visible tenseness, closed fist shaking menacingly, a coiled-spring look, and growling tone of voice all convey a forceful neuromuscularly stated message.

The description just offered makes a furious man seem very like an enraged beast, and, of course, for over 110 years ethologists — Darwin (1872) was undoubtedly the most thorough and reliable of the nineteenth century pioneers — have noticed and documented the striking consonances between the patterned emotional activities of animals and those of man. More recently, Ekman and colleagues (Ekman & Friesen, 1975; Ekman et al., 1983; Ekman, 1984) have established the cross-cultural similarities of human affect-specific facial expressions, and MacLean's (1952, 1967, 1969, 1972, 1977, 1978) explorations in neuroevolution and neuroethology have led him to propose that even quite civilized peoples show a number of the fixed action patterns seen not only in primates but also in lower vertebrates. Again, although perhaps originally the products of repeated and accidental genetic mutations, these behavioral stereotypies have endured probably because of their functional and adaptive value: the motor features of the anger response, for example, prime the creature to spring and bite.

In experimental animals, electrophysiologic stimulation of various limbic system and brainstem loci can evoke the characteristic motor and hypothalamic-autonomic responses which "inscribe" these visible and audible "signatures" of specific emotions —

findings underscoring the "hard-wired" nature of these stereotypies (for reviews, see MacLean, 1952, 1967; Kupferman,
1985a,b)—and in man, moreover, the capacity to generate and
decipher the patterned gestures, inflections, and timbres that
constitute *prosody* and give speech its affective overtones apparently demands the integrity of and contributions of frontal and
temporal cortical elements contralateral to and possibly homologous to the areas of Broca and Wernicke necessary for verbal
expression and comprehension (Heilman et al., 1975; Ross, 1984;
Kandel, 1985a).

Evidently, then, and in a manner again comparable to that of
the classic senses, the raw afferent stimuli carrying emotional
messages undergo processing essentially as stereotyped as is the
retina's and visual pathway's handling of a bar of light flashed on
a darkened screen (Hubel, 1982). More specifically, data gathered
by conventional sensory systems travel via shunts (cf. Castellucci,
1985) into circuitry subserving emotion and there apparently
submit to an as yet undefined "extraction procedure" which yields
the "refined" affective "product"—for example, the subjective
"startle" discussed below.

All afferent data, then, receive *parallel processing* by both
sensory and affect-generating circuits, and the acoustic startle-
response illustrates not only this two-channel assessment principle
but also its immediate relevance to a comprehensive understanding
of emotional communication. When we jump and spin around on
hearing a sudden, unanticipated noise, our auditory system provides detailed information about the direction, source, loudness,
and additional sonic properties of the stimulus, but the rush of
anxiety and tachycardia that we might well note issue from the
participation of limbic, right temporal cortical, and hypothalamic-
autonomic components in the appraisal of the sound.

In sum, then, while much pertinent neurobiologic detail obviously awaits experimental elaboration, we can with safety conclude that, because of our sophisticated neural audio-analyzers
and prosody-sensitive circuitry of the right temporal lobe, we
decode   rather   automatically   and   accurately   the   nonverbal

"metacommunicational" (Jacobs, 1986) messages that our fellow humans—and animals—transmit in equally stereotypic fashion. If, for instance, someone speaks harshly to us in the tone we label contemptuous, we react with felt anxiety, shame perhaps, and even depression; and we know from vocal timbre but without thinking whether our superior has told us "good job" as a compliment or as an insult. One nearly need not add that such "unconscious mechanisms" unquestionably influence transference-countertransference exchanges.

For the moment, however, with a view to defining the relevance to technique of the preceding exposition, one might propose that *if indeed each emotion has its distinctive neuromuscular and hypothalamic-autonomic "signature," then by recognizing the clinically evident stereotypies, the analyst can with some precision identify which affects are currently close to or at "the surface."* A change in tone of voice, a sudden hushed quality, for example, can cue the listener to the patient's subjective surge of, in this case, embarrassment.

At this point, however, some words of amplification might prove useful: to imply that tone of voice is a motor act may not seem an obviously logical step. Nevertheless, just as a wind instrument's sound derives from, among other properties and variables, speaking length, tube shape, scaling, resonator type, wall compliance, and wind pressure, so do the pitch, timbre, and loudness of human utterances depend on comparable, even identical factors controlled not, of course, by a brassmith's hands or hornist's technique, but rather by neural circuits that automatically adjust tonus and coordinate contraction within laryngeal, pharyngeal, palatal, lingual, labial, buccal, diaphragmatic, and intercostal muscle groups. Naturally, one can hear more than "muscular" signs of emotion, and hypothalamic-autonomic outflow also adds feeling-specific contributions to aurally detectable data: a patient's sudden nasality may point to tears that have accompanied a rush of sadness, for example, and a rasping harshness can indicate an anxiety that has led to inhibition of the normal secretions of larynx and pharynx.

Affect-specific audible and visible data from the analysand are not, though, the only neurobiologically generated phenomena to which a clinician might carefully attend, and, as suggested above, the analyst's own subjective feelings often, because of "hard-wired" and stereotyped ways of processing "transmissions" from other people, significantly reflect affective themes within the patient—emotional currents whose overt manifestations may initially have escaped notice. However, as the following sections will emphasize, there exists no single and simple correspondence between the experiences of the treatment's two participants: withholding of words, fees, or attendance at sessions can, for instance, anger a therapist, but the provocative behaviors could in fact derive from stubborn rage, paralyzing fear, immobilizing depression, excruciating shame, or some variable mélange of the four. Awareness of countertransference annoyance, obviously, would not alone provide precise and trustworthy clues to an interpretable "affective surface."

Nonetheless, knowledge of and diligence in observing the sometimes subtle signs of affect-triggered stereotypies arguably supply the "reagents" for a useful and reliable "in-the-hour-lab-test" for emotion, and thus, one should perhaps conclude, that ostensibly ineffable talent at times called "empathy" may reside ultimately in refinable skills at recognizing definable and really quite physical data.

"Son et Lumière": Sounds, Sights, and Other Signs of Individual Emotions

If, then, "empathy" to a degree begins with a sort of "neurological examination"—the apperception of stereotypic motor and autonomic signs of affect—one might at this juncture survey the emotions of clinical importance and try to cull their characteristic manifestations. Influenced and informed by works of several authors (e.g., Darwin, 1872; Dahl, 1978; Ekman & Friesen, 1975; Ekman et al., 1983; Knapp, 1981, 1987; Plutchik, 1980; Tomkins, 1970), the classification to follow copies none, borrows to an extent from each, and generally takes its own path toward defining broad categories of feeling, all of which include a considerable spectrum of intensity.

The emotions of the *anxiety/fear* class can, obviously enough, range in urgency from mild "nervousness" to utter terror, and its more microscopic neuromuscular accompaniments may include startle reactions and subtle tremor of hands, feet, voice, and even facial structures—oscillations of muscle contraction presumably of extrapyramidal origin and thus conceivably involving altered dopaminergic neurotransmission. In addition, one not infrequently observes such larger-scale stereotypies as raising an arm to shield the head—one sometimes sees this response in patients on the couch; flight from frightening situations—or thoughts; and paralysis or "freezing" stemming from dread.

Furthermore, as indicated earlier, anxiety commonly betrays its presence through hypothalamic-autonomic reactions—perspiration, piloerection perhaps, and the raspiness of voice that often follows inhibition of normal glandular secretions in the larynx and pharynx—and while all these phenomena may aid the clinician's attempts to read his patient, the last of the three, one must add, is one nonverbal sign that can alert the *analysand* to unusual discomfort in the unseen person behind the couch.

In line with preceding discussions of stereotyped emotional "data-processing," one should note the several kinds of sensory information that seem to evoke anxiety in rather "hard-wired" fashion. As already described, sudden, loud, or harsh sounds can trigger the startle response and its affective accompaniment, and in addition, fast-approaching images in the visual fields may elicit both a flinch and a fearlike feeling.

Furthermore, and probably via the earlier mentioned affect-decoding circuits of the right temporal lobe, such prosodic properties as angry, snarling, and contemptuous vocal timbres and corresponding facial expressions powerfully stir anxiety. Moreover, plausibly through comparable neural mechanisms, signs of uneasiness or fear in one person can evoke these feelings in others—someone who is shaking like a leaf makes everyone

uncomfortable—and this phenomenon exemplifies well the apparently "hard-wired" manner in which the overt manifestations of an
emotion in one individual will often elicit the same sensation in
people nearby. Since Darwin's time, ethologists have described
this sort of affective "contagion," and this form of communication
obviously confers considerable evolutionary and adaptive advantage: by promoting, for example, the rapid spread among herd
members of an awareness of danger and dread, this "inter-animal
shorthand" may have immediate survival value. This automatic
"information transfer" probably underlies the emotional "resonance" that we include under the rubric "empathy."

A further and significant form of anxiety-producing paradigm
not only proves easily demonstrable in nonhuman vertebrates but
also intrudes fairly regularly into commonplace exchanges both of
parent and child and of analyst and patient. More explicitly, if one
picks up an ordinary laboratory rat gently but from the back, the
rodent will often struggle agitatedly, kick his legs, and try to turn
around and bite. Similarly, if one first operantly conditions a
squirrel monkey to press a bar five times to obtain a food reward
and then suddenly discontinues the reinforcement, the little pri-
mate will frequently shriek, thrash frantically, and attempt to nip
whomever he might. In both instances, what one may conclude is
that when the organism finds that familiar muscular efforts no
longer accomplish the results experience has led it to expect,
panic—and rage—follow. Quite arguably, the fundamental mechanistic principle is that an animal, on a "hard-wired" basis, will
react with intense alarm if he discovers that he has lost motor
control of his body's comings, goings, and doings. Nor is it too
difficult to infer the evolutionary and adaptive value of such a
response: it's a warning that informs the creature that he's
"spinning" his neuromuscular "wheels."

Obviously, this type of reaction could well underlie the "autonomy" struggles that at times infect parent-child interactions
around feeding, locomotion, and toilet training. Furthermore, as
hinted above, homologous issues can contaminate also the psychoanalytic situation. Specifically, a patient may feel panicked

and frantic if he believes he must swallow, follow, or submit to an intervention that, because of errors of focus, timing, or tact, fails to find an acknowledgeable "affective surface." Similarly, an analyst might experience comparable emotions of helplessness if he concludes that the analysand and the therapeutic process are escaping his efforts at interpretive management — a significant countertransference predicament.

Turning now to the category of *sadness/depression,* one observes again a considerable range of intensity and duration — brief moments of "hurt feelings," for example, arguably consist of a transient, sharp "spike" of depressive pain associated with some feature of "self representation" (Schwartz, 1990a) — and we here encounter also rather familiar motoric hallmarks — hanging head, retarded movement and speech tempo, drooping facial features, frown, and downcast eyes. Apart from sleep and appetite disturbance, "vegetative signs" that are only reportable but not, obviously, audible or visible in the consulting room, the salient hypothalamic-autonomic response is that of lacrimation, a phenomenon one can at times first note aurally because tears flow through the nasal-lacrimal duct, change the resonating properties of the upper nasal chambers, and give the voice a telltale "stuffy nose" sound.

Sadness/depression seems fundamentally a response to loss — of person, sustenance, or self-esteem (cf. Brenner, 1982) — and, of course, the deprivation may involve not only what one has had in hand but also what one had expected: a "B" on an exam can feel devastating to one who had anticipated something better. Furthermore, as with anxiety, angry or scornful vocal tone can evoke depressed feelings, and, once more, "contagion" can "spread" this affect from one individual to another, probably again contributing to the "affect-sharing" experiences termed *empathic resonance.*

Although perhaps more properly considered a sensation rather than an affect — and at any rate illustrating nicely the somewhat artificial distinction we draw between the two classes of phenomena — *disgust* nonetheless, because of its potentially intense and powerful learned associations with sexual or excretory anatomy and function, with eating and images of food, and with mental representations of aspects of self and others, has notable clinical importance. Physiologically, this feeling triggers programmed motor and hypothalamic responses preparatory to vomiting, and one can often hear a "yuck" sound in the voice and see facial contortion when this state intensifies subjectively. Although obviously not apparent to the observing analyst, increased salivation and palpably reversed peristalsis are among the associated parasympathetically-mediated concomitants that patients may report.

Both as Freud (1905) noted and as suggested above, learning can associatively link disgust to a variety of other phenomena, which will then acquire, as Freud observed regarding feces, the ability to call forth feelings of distaste, and, in addition, certain smells, tastes, and textures, particularly slimy consistencies, appear in a more "hard-wired" manner to evoke this unpleasant sensation. Furthermore, the sounds and sights of nausea and vomiting seem to have an automatic and "contagion"-like power to elicit similar reactions in others; thus parents' facial and vocal expression of "disgusted" responses may well serve as unconditioned stimuli in teaching children, via a Pavlovian paradigm, to share their elders' values (Schwartz, 1987, 1988, 1990a; cf. Reiser, 1984, 1985; Rescorla, 1988).

Because neither researchers nor clinicians have yet established a clear and convincing separation between *embarrassment* and *shame,* one may perhaps usefully view these terms as signifiers of a common class of affect that differ primarily in the greater intensity and noxiousness that the latter word implies (cf. Brenner, 1979). Dann (1977) concludes that Freud did not distinguish linguistically between the two nouns in question, and that observation might for the moment suspend debate until further and possibly differentiating data arrive from primary investigators (cf. Ekman et al., 1983).

At any rate, the feelings to which the words *embarrassment/ shame* apply evoke in someone pained urges to hide, and these emotions' repertoire of characteristic motor stereotypies includes several avoidant actions: one bows the head, one shields the face with the hands, one lowers the eyelids and averts the gaze to escape the intimate "discovery" which direct eye contact threatens, and one finally slinks off. Moreover, the voice frequently has a rather easily identified breathy hush: one also "covers up" speech. Obviously, too, a visible autonomic reaction may become apparent — a blush.

Evoking embarrassment/shame, one might propose, are those experiences in which one feels oneself the object of contempt, scorn, laughter, or derision, and MacLean's ethologic observations, however anecdotal, nevertheless seem intringuingly consistent both with the suggestion offered here and with Darwin's similar inferences. Specifically, when animals confront each other over issues of territory, social precedence, or sexual competition, they engage in display and threat behaviors — and, in a manner with, arguably, a clear homologue in human sarcasm and contumely, they growl and snarl at one another. Additionally, and more pertinently, if one combatant successfully "out-snarls" the other, the defeated beast will put tail between legs, lower head, sink toward the ground, and grovel off. MacLean notes such fixed action patterns in a species of lizard; he reports that the vanquished reptile may undergo profound hypothalamic-autonomic changes that can terminate in death — in literal "mortification."

Finally, emotional "contagion" seems again capable of eliciting in onlookers the embarrassment or shame detectable in another: we squirm and bow our heads at the humiliation of our comrade.

Like the emotions already discussed, *anger/contempt* exhibits a considerable spectrum of intensity — from mild annoyance to blind and ineffable fury — and an earlier section has described the motor stereotypies which this variety of affect may trigger. One might note again and even stress, though, that the vocal snarl which we know familiarly as sarcasm, scorn, and contempt is a frequent manifestation of this feeling — and a lamentably common sign of malignant countertransference enactment, too, one should add (Jacobs, 1986). This timbre, as suggested above, can automatically stir in the intended listener a potent cocktail of anxiety, depression, humiliation—and rage itself.

This last point merits emphasis. Fundamentally, neurobiologic, ethologic, experimental psychologic, and even clinical studies (Darwin, 1872; MacLean, 1952, 1969; Adamec, 1975; Holt, 1976; Kupferman, 1985a; Schwartz, 1987; Stechler & Halton, 1987) yield data most consonant with the conclusion that anger, its variants, and its associated neuromuscular stereotypies are not manifestations of an appetitive "drive" but rather of a genetically "blue-printed" affectomotor response to threat—of, in other terms, an element of our inborn repertoire of defenses. Thus, when the organism feels challenged or endangered and therefore experiences a dysphoric affect of anxiety, depression, or shame, it will very likely react also with hostility. Humans frequently display violence of greatest intensity after having suffered a galling mortification—and we all know what to look out for in a wounded animal.

While anger/contempt may also impress observers as potentially "contagious," the mechanism of spread in this case probably differs significantly from the more direct "contagion" discussed on several occasions above. If, as mentioned earlier, one is speaking with an individual who fidgets, squirms, and trembles visibly, one feels uncomfortable—one has "caught it," so to speak—but if, on the other hand, an analyst is listening to a patient who is screaming at him, branding him with every known slur, and sarcastically taunting him with each of his perceived character flaws and technical lapses, the therapist might indeed feel furious, too—though not now out of "empathic resonance." Instead, this rage would reflect a reactive response to the helpless anxiety and humiliation which the abuse has induced; and one sees not "shared experience" but raw wishes for revenge—the emergence of the "talion principle" which Racker (1957) so stresses. Finally, as suggested elsewhere in greater detail (Schwartz, 1987), this kind of relatively "hard-wired" affective communication can perhaps account for the complex interpersonal provocations and influences which Klein and her colleagues (Segal, 1964) attempt to explicate with their recondite formulations of projective identification: if one appreciates how such sonic properties as timbre and loudness affect human ears and emotions, one need no longer invoke "projected introjects" to explain many intricacies of our species' mutual interactions.

As detailed earlier, the five varieties of affect just discussed—anxiety/fear, sadness/depression, disgust, embarrassment/shame, and anger/contempt—all represent highly evolved and complex psychobiological alarm reactions which share a further psychologically significant characteristic: they evoke behaviors, actions both overt and "in the head," that are protective and defensive and that serve to rid oneself or one's world of the stimuli that evoked the feelings in the first place. One trembles before and flees from what terrifies; hides or slinks from mortifying thought or opponent; vomits the "yucky"; elicits help and sustenance from fellow beasts if one shows the sluggishness and lethargy of loss or bereavement; and escapes the oppression of the once-dominant but at-last-defeated rival.

Furthermore, the first four dysphoric emotions listed above can, if associatively linked to mental representations of feeling, impulse, self, or other, serve as the "signal affects" (Freud, 1926; cf. Schwartz, 1987, 1988, 1990a) which initiate escape or defensive strivings. Obviously, too—and this point has notable technical implications—each "alarm" is often clinically discernible, either through its characteristic motor stereotypies or through the subjective responses it may evoke.

At this juncture, one might pause to note briefly three emotional phenomena whose absence from the preceding exposition may lend them a conspicuousness even greater than their typically prominent place among clinically critical affective issues. One can argue, however, that *envy, jealousy,* and *guilt* do not represent pure cultures of feeling but rather constitute blends of primary emotions.

Envy, for example, comprises, one might suggest, depression, anxiety, anger certainly, and, not least, significant strains of humiliation—all experienced in a setting of thoughts that, at base, carry the theme that "He (or she) has what I lack." Jealousy, moreover, seems a similar mélange—but one with a somewhat different ideational context: this emotional mixture generally takes as target two people and their exclusive and excluding relationship with each other.

Guilt, too, one can propose, comparably constitutes a combination of feelings that mixes anxiety, depression, and anger directed at oneself—the three now experienced along with thoughts of "What have I done!"—or failed to do. Furthermore, this blend may include also a dash of the empathically felt pain of the victim of one's abuse or neglect.

While focus has so far rested on dysphoric feelings, one must of course also consider hedonic affects and sensations which, as detailed earlier in this paper, serve as compelling goals and powerful reinforcers of learned emotional behavior.

Although Freud (1914) postulated a common "libidinal" source of the excitement he observed both in sexuality and in grandiosity, the researchers cited above suggest strongly, as argued elsewhere (Schwartz, 1990a), that the brain-generated euphoria seen in drug-induced and manic intoxications, the emotional product of dopaminergic and opioid-rich mesolimbic circuits, is a species of pleasure quite distinct from, though at times synergistic with, the sensual gratifications of erotic arousal and orgasm. As proposed earlier (Schwartz, 1990a), one may now draw the conclusion that this affective "high" functions as the principal endogenous reward which signals achievement of a host of survival-critical tasks—the acquisition of territory, social power, and precedence; the attracting of appealing, fertile, responsive, and receptive mates; the finding of impressive and protective leaders; and the conquest of threat and complexity.

As have the dysphoric emotions, euphoria also has its characteristic neuromuscular stereotypies—bouncing walk; rapid, high-pitched speech; broad smile; laughter—and, additionally, related feelings called "wonder" and "awe" can lead one's jaw to drop or make one go a bit limp all over—as one may if one hears a marvelous sound, tastes an exquisite sauce, or sees a gorgeous building or face—experiences not here, at any rate, appearing in rank order of aesthetic appeal. Closely associated with affective "high" are rituals of exhibitionism: animals and man strut; flaunt antlers, feathers, or clothes; puff out their chests; and even, as in the case of primates, display buttocks and genitals. Obviously in ways suggesting sexuality, these patterned activities also occur in the context of competition for territory and social precedence.

Again as with dysphoric feelings, certain stereotyped forms of sensory stimulation can evoke euphoria, and both infant researchers and observant parents know that neonates will show signs of delight if one smiles, coos, or squeaks at them—responses that are evidently "hard-wired." Laughter, too, seems "infectious"—witness the canned laugh track of television situation comedies. Moreover, as an earlier essay (Schwartz, 1990a) proposes in detail, patterned regularities and symmetries of visual and auditory input seem to elicit affective "high" and may thus underlie common aesthetic preferences.

The foregoing discussion introduces the notion that euphoric and sexual pleasures represent distinguishable, if often intertwined, sensory/affective phenomena, and an earlier article (Schwartz, 1990a) develops the further hypothesis that human sexuality gains its considerable and species-survival essential appetitive appeal from the synergistic contributions of at least two anatomically and neurobiologically separable reward circuits— one generating a "high" and triggered by information conveying the interest and receptivity of an attractive potential mate, the other yielding the different but also exquisite gratifications of erotic excitement and orgasm.

Therefore, with the aid of current concepts and data, one might suggest that sexuality seems not a "drive" or "instinct" propelled by tissue deficit or fueled by "libidinal energy" but rather a complex sensorimotor program activated by specific stimuli and appetitively pursued because of the intense pleasures generated (cf. Holt, 1976; Klein, 1976; Kupfermann, 1985a, 1985b). In different words, one can propose that we vertebrates have evolved in such a fashion that our neural circuitry rewards us richly if we complete an act critical for species survival—and, one hardly need add, we seek to repeat this "genetically-blueprinted" routine in order to reexperience the gratifications afforded.

Sexual interest and stimulation also initiate stereotypic actions; that repertoire of behaviors which we call "flirting" includes activities with homologues at many levels of the evolutionary tree: animals and humans look over and at prospective mates, sniff one another, show off and display themselves, and make playful attempts to dominate each other with seductive teasing. Clinically, one sees examples of most of these manifestations, and one may additionally note audible signs of romantic and erotic preoccupation and desire—for instance, deep sighs and the husky, throaty tone of voice one sometimes hears in patients with "hysterical" character traits.

Considering the stimuli that spark romantic and sexual arousal, one can observe that all of the flirtatious activities noted above seem potentially effective in soliciting the interest of their intended target. Visual behaviors are perhaps particularly evocative—for example, the proverbial soulful glance at the person for whom one longs—and for a love-besotted individual, there are probably few, if any, more powerful intoxicants than a gaze returned, eyes looking directly into eyes.

Obviously, chemical and physical stimuli can stir or augment romantic or erotic interest—thus the importance of perfumes and of those aesthetically-pleasing facial and bodily configurations labeled "beauty"—and one might add, many personal attributes evoking euphoric "high," "special" qualities that can range from physical appeal to political power to wealth, also tend, as suggested elsewhere (Schwartz, 1990a) in somewhat greater detail, to elicit sexual yearnings.

As hinted a bit earlier, perhaps more than two sensory/affective reward mechanisms participate in human sexual and interpersonal behaviors. Sandler and Joffe (1968, 1969) and Rosenblatt (personal communication, 1987) emphasize the importance of the low intensity yet warmly pleasant emotion which Sandler and Joffe (1968) label the "safety feeling." Whether this hypothesized affect is indeed a totally distinct entity may remain debatable, but a soft, anxiety-free, and deeply rich tone of voice can suggest that a patient is experiencing the general sort of state Sandler and Joffe describe.

Not Art But Science: Biologic Analysis and Interpretive "What's" and "When's"

Correct focus, timing, and tact, a majority of analysts could conceivably agree, constitute perhaps the most crucial elements of a successful interpretation. One might at this juncture examine how the concepts and data so far presented may aid the clinician in deciding what to address and when to say it.

What one might most stress regarding selection of interpretive *focus* probably seems now to have been obvious, if implicit, in the preceding expositions: *if one attends carefully to the detectable and characteristic nonverbal signs specific for each affect, one has immediately useful clues to which feelings are—at that moment—close to or "at the surface," and one can thus with considerably less difficulty choose and formulate the proper "emotional plot" for an upcoming intervention.*

Furthermore, as hinted at this paper's beginning, by comparing systematically—and before delivery—an interpretation's explicit or covert affective syntax and logic with the "hard" physical data the patient generates, one can more rationally determine whether the intervention's *emotional* meaning meshes with the analysand's current experience, and one thus has a genuinely scientific method for *predicting* if a comment will land "on target"—or ricochet off the wall.

For example, as many analysts realize, remarks intended as "defense interpretations" not infrequently strike patients as criticisms, and analysands then fundamentally feel as if they're receiving the message, "You're avoiding what you *should* be talking about. You're eating dessert before your peas. *Naughty."* Underlying this fairly common occurrence, one might suggest, is a failure to key the comment to the detectable, acknowledgeable presence of an unpleasant emotion which sparks, for instance, a sudden and protective change of topic; the intervention may instead reflect inferences drawn in turn from "verbal material," often a less trustworthy guide to conscious and preconscious *affective* experience.

Painstaking attention to nonverbal signs of affect, on the other hand, can make "defense interpretation" a much more scientific, reliable, and—for the patient—palatable and useful enterprise. More particularly, if the analyst succeeds in identifying and judging acknowledgeable a dysphoric feeling that seems to trigger a defensive shift of subject matter, then the clinician may with notably less guesswork formulate a comment to which the analysand will respond, "That's right. I didn't really realize it—but I am too scared"—or embarrassed or guilty or whatever—"to talk with you further about *that* topic." In other words, when an intervention addresses directly or at least implies emotion that the listener actually experiences, then the chances increase that the remark will convey valuable information—and not provoke bewilderment or fear, hurt or humiliation.

Of course, study of motor stereotypies is not the only biologically-aided access we have to the interpretively accessible "affective surface," and, as discussed earlier, the automatic and evidently "hard-wired" processes of "emotional contagion" may evoke in the "resonant" clinician the anxiety, embarrassment, disgust, or depression his patient experiences. On the other hand, as already emphasized, a therapist's subjective reactions have no uniform or facile translation into accurate readings of the feelings of the other party in the office. A careful comparison of one's own affects with nonverbal and prosodic data from the analysand should help protect against the too-easy assumption that what the analyst tastes pertains also to the person on the couch.

A further path to the emotional "surface" leads through creative use of the imagination (Buie, 1981) and depends on our wondering silently how we ourselves would feel were we in the situation which the patient describes or evokes. However, some authorities caution sternly against this sort of introspective exercise and quite defensibly argue that idiosyncrasies of personal history may engender significant misreadings: on the one hand, the analysand has perhaps had unusual experiences very unfamiliar to the therapist, and, on the other, the clinician's own life events can, so to speak, have inflexibly "prefocused" or even warped his "empathic lenses." A senior analyst tells of the surprising response he encountered to an interpretation that assumed a love of the exciting and prestigious university that he and the young woman on the couch had both attended: the lady had in fact hated the place.

Obviously, this sort of error can and does occur, but if, on the other hand, one uses available evidence of nonverbal clues to affect to set, so to speak, the scene of one's imaginings, one will more likely attain a reliable subjective inference. Should one, for instance, note that the analysand characterizes an employer's voice as biting, one readily senses how hearing that snarl would feel. The key fact that affect-related information processing is rather stereotyped ought give one some confidence that, when one considers carefully each hint of emotion from every participant in the situation described, one may well find that introspective efforts to put oneself in the patient's place do yield useful information.

Furthermore, and one might stress this idea, one can—and should—use *multiple but converging lines of evidence* to "diagnose" what feelings are "at the surface." Prosodic and other nonverbal signs of emotion, subjective affective responses, and "trial by imagination" all provide valuable clues, and one may check the implications of a given set of data against the inferences one draws from the remaining varieties of information. A thorough and competent internist will base his working impressions on the *combined* yield of history, physical examination, blood studies, and radiologic findings: he puts his patient—and himself—at risk if he chooses to rely on a sole source of guidance.

Implicit in the preceding discussion of interpretive focus is an assumption quite relevant also to a scientific consideration of *timing*—that the *acknowledgeability* of an affect, and thus the likelihood that an analysand will say "Right! Exactly!" to an intervention naming or implying that feeling, *varies in direct proportion to the amplitude, intensity, duration, or frequency of the neuromuscular or autonomic signs of that emotion.* Obviously, this latter proposition is just an author's personally validated inference which for now other practitioners can only put to the test of their own experience, but what one might underscore is that this idea is also an hypothesis potentially amenable to validation or refutation via such quantitative techniques as measurement of parameters of sympathetic and parasympathetic activity or, conceivably, power-spectrum analysis of voice frequencies (cf. Ekman et al., 1983; Ekman, 1984).

If one agrees or confirms that, for example, the faster the patient's foot shakes, the more intense and thus acknowledgeable his anxiety, then one can propose that, in principle, timing can become fundamentally a rather simple matter: one intervenes at the crescendi of affect, at the emotional accents — sometimes relatively subtle — which one gauges almost quantitatively from prosodic and other nonverbal data. Half a century or more ago, Klein (Strachey, 1934), Strachey (1934) himself, and Fenichel (1941) all in effect advised that one interpret at points of "urgency" (Strachey, 1934), and today one might only add that refined neuroscientific and ethologic concepts and observations may abet our locating the *marcato* passages.

What one might in sum suggest is that we are now more able to operationalize "empathy," to dissect the sometimes mystical-seeming business of reading people, and thus to determine which data and information-gathering processes appear most crucially involved. Specifically, we can begin to define the discrete abilities on which we depend — for instance, that of recognizing signs of different affects in tone of voice — and a skill identified is one that may improve with focused and disciplined practice: concentrated effort should sharpen our listening for the changes of timbre that indicate disgust or anxiety, embarrassment or euphoria.

Obviously, the exposition above suggests strongly that the vocal qualities which inform us of the presence and intensity of emotion are essentially musical properties — timbre, pitch, rhythm, and, naturally, loudness — and the practice of psychoanalysis therefore involves, as Greenson (1967) noted, the skills and sensitivities one uses in, for instance, listening to a concerto. Certain genetically programmed and thus heritably variable gifts undoubtedly help shape our clinical abilities — one thinks immediately of acuteness of tonal discrimination — but again, one must stress that inborn talents usually improve with intelligent practice: a pocket score, a musicassette, and a "Walkman" can significantly aid one's learning to recognize the members and combinations of the orchestral wind choir.

Many, perhaps most, experienced clinicians might now suggest that matters of interpretive focus and timing show complexities far greater than those of the relatively simpler challenges of detecting the presence and accents of motivationally significant affects. One could argue that what a patient feels an analyst cannot always successfully address, and one would perhaps cite as an example an analysand stuck in the crater of volcanic transference rage and able to do no more than throw forth hot chunks of fury. In such an instance, the emotion in question is probably apparent enough to all interested parties — but trying to talk about the anger with the person experiencing it can sometimes resemble an attempt to put out a gasoline fire with acetone.

However, the fact that an affect whose presence one infers from nonverbal data can nonetheless *appear* "beyond interpretation" does not imply that the emotion is neither conscious nor preconscious. What may in fact be at issue is the *manner* in which the analyst tries to address the feeling he detects, and one thus veers toward the complexities of *tact,* neuroscientific approaches to which topic will be a principal focus of a future essay (Schwartz, in prep.).

REFERENCES

Abend, S. M. (1986). Some problems in the evaluation of the psychoanalytic process. In: *Psychoanalysis. The Science of Mental Conflict. Essays in Honor of Charles Brenner,* ed. A. D. Richards & M. S. Willick. Hillsdale, NJ: The Analytic Press, pp. 209–228.

Adamec, R. (1975). Behavioral and epileptic determinants of predatory attack behavior in the cat. *Canadian J. Neurological Sciences,* 457–466.

Allon, N., & Wollberg, Z. (1978). Responses of cells in the superior colliculus of the squirrel monkey to auditory stimuli. *Brain Research,* 159:321–330.

Arnold, M. B. (1970). Perennial problems in the field of emotion. In: *Feelings and Emotions: The Loyola Symposium,* ed. M. B. Arnold. New York: Academic Press, pp. 169–185.

Brenner, C. (1976). *Psychoanalytic Technique and Psychic Conflict.* New York: Int. Univ. Press.

———— (1979). Depressive affect, anxiety, and psychic conflict in the phallic-oedipal phase. *Psychoanal. Q.,* 48:177–197.

———— (1982). *The Mind in Conflict.* New York: Int. Univ. Press.

———— (1987). A structural theory perspective. *Psychoanal. Inquiry,* 7:167–171.

Buie, D. H. (1981). Empathy: Its nature and limitations. *J. Amer. Psychoanal. Assn.,* 29:281–307.

Castellucci, V. F. (1985). The chemical senses: Taste and smell. In: *Principles of Neural Science,* 2nd ed., ed. E. R. Kandel & J. H. Schwartz. New York, Amsterdam & Oxford: Elsevier, pp. 407–425.

Dahl, H. (1978). A new psychoanalytic model of motivation: Emotions as appetites and messages. *Psychoanal. & Contemp. Thought,* 1:373–408.

Dann, O. T. (1977). A case study of embarrassment. *J. Amer. Psychoanal. Assn.,* 25:453–470.

Darwin, C. (1872). *The Expression of the Emotions in Man and Animals.* Chicago: Univ. Chicago Press, 1965.

Ekman, P. (1984). Expression and the nature of emotion. In *Approaches to Emotion,* ed. K. Scherer & P. Ekman. Hillsdale, NJ: Lawrence Erlbaum, pp. 319–343.

———— & Friesen, W. V. (1975). *Unmasking the Face.* Englewood Cliffs, NJ: Prentice-Hall.

———— Levenson, R. W. & Friesen, W. V. (1983). Autonomic nervous system activity distinguishes among emotions. *Science,* 221:1208–1210.

Fenichel, O. (1941). *Problems of Psychoanalytic Technique.* New York: The Psychoanalytic Quarterly Inc.

Freud, S. (1895). Project for a scientific psychology. *S.E.,* 1.

———— (1905). Three essays on the theory of sexuality. *S.E.,* 7.

———— (1914). On narcissism: An introduction. *S.E.,* 14.

———— (1926). Inhibitions, symptoms, and anxiety. *S.E.,* 20.

Gerig, R. R. (1974). *Famous Pianists and Their Technique.* Bridgeport, CT: Luce.

Gill, M. M. (1982). *The Analysis of Transference, Vol. I.* New York: Int. Univ. Press.

Gray, P. (1982). "Developmental lag" in the evolution of technique for psychoanalysis of neurotic conflict. *J. Amer. Psychoanal. Assn.,* 30:621–655.

———— (1986). On helping analysands observe intrapsychic activity. In: *Psychoanalysis. The Science of Mental Conflict. Essays in Honor of Charles Brenner,* ed. A. D. Richards & M. S. Willick. Hillsdale, NJ: The Analytic Press, pp. 245–262.

Greenson, R. R. (1967). *The Technique and Practice of Psychoanalysis, Vol. I.* New York: Int. Univ. Press.

Heilman, K. M., Scholes, R. & Wilson, R. T. (1975). Auditory affective agnosia. *J. Neurol. Neurosurgery, & Psychiat.,* 38:69–72.

Holt, R. R. (1976). Drive or wish? A reconsideration of the psychoanalytic theory of motivation. In: Psychology versus metapsychology: Psychoanalytic essays in memory of George S. Klein, ed. M. M. Gill & P. S. Holzman. *Psychol. Issues,* Monogr. 36. New York: Int. Univ. Press, pp. 158–197.

Hubel, D. H. (1982). Exploration of the primary visual cortex, 1955-78. *Nature,* 299:515–524.

Jacobs, T. J. (1986). On countertransference enactments. *J. Amer. Psychoanal. Assn.,* 34:289–307.

Kandel, E. R. (1985a). Brain and behavior. In: *Principles of Neural Science,* 2nd ed., ed. E. R. Kandel & J. H. Schwartz. New York, Amsterdam & Oxford: Elsevier, pp. 3–12.

_____ (1985b). Processing of form and movement in the visual system. In: *Principles of Neural Science,* 2nd ed., ed. E. R. Kandel & J. H. Schwartz. New York, Amsterdam & Oxford: Elsevier, pp. 366–383.

Kelly, J. P. (1985). Auditory system. In: *Principles of Neural Science,* 2nd ed., ed. E. R. Kandel & J. H. Schwartz. New York, Amsterdam & Oxford: Elsevier, pp. 396–408.

Klein, G. S. (1976). Freud's two theories of sexuality. In: Psychology versus metapsychology: Psychoanalytic essays in memory of George S. Kelin, ed. M. M. Gill & P. S. Holzman. *Psychol. Issues,* Monogr. 36. New York: Int. Univ. Press, pp. 14–70.

Knapp, P. H. (1981). Core processes in the organization of emotions. *J. Amer. Acad. Psychoanal.,* 9:415–434.

_____ (1987). Some contemporary contributions to the study of emotions. *J. Amer. Psychoanal. Assn.,* 35:205–248.

Kupfermann, I. (1985a). Hypothalamus and limbic system I: peptidergic neurons, homeostasis, and emotional behavior. In: *Principles of Neural Science,* 2nd ed., ed. E. R. Kandel & J. H. Schwartz. New York, Amsterdam & Oxford: Elsevier, pp. 611–625.

_____ (1985b). Hypothalamus and limbic system II: motivation. In: *Principles of Neural Science,* 2nd ed., ed. E. R. Kandel & J. H. Schwartz. New York, Amsterdam & Oxford: Elsevier, pp. 626–635.

_____ (1985c). Genetic determinants of behavior. In: *Principles of Neural Science,* 2nd ed., ed. E. R. Kandel & J. H. Schwartz. New York, Amsterdam & Oxford: Elsevier, pp. 795–804.

MacLean, P. D. (1952). Some psychiatric implications of physiological studies on frontotemporal portion of limbic system (visceral brain). *Electroenceph. Clin. Neurophysiol.,* 4:407–418.

_____ (1967). The brain in relation to empathy and medical education. *J. Nerv. Ment. Dis.,* 144:374–382.

_____ (1969). The hypothalamus and emotional behavior. In: *The Hypothalamus,* ed. W. Haymaker, E. Anderson & W. J. H. Nauta. Springfield, IL: Charles C Thomas, pp. 659–678.

_____ (1972). Cerebral evolution and emotional processes: new findings on the striatal complex. *Annals N. Y. Acad. Sci.,* 193:137–149.

_____ (1977). On the evolution of three mentalities. In: *New Dimensions In Psychiatry: A World View. Vol. 2,* ed. S. Arieti & G. Chrzanowski. New York: Wiley, pp. 306–327.

_____ (1978). A mind of three minds: Educating the triune brain. In: *Seventy-seventh Yearbook of the National Society for the Study of Education.* Chicago: Univ. Chicago Press, pp. 308–342.

Martin, J. H. (1985). Receptor physiology and submodality coding in the somatic sensory system. In: *Principles of Neural Science,* 2nd ed., ed. E. R. Kandel & J. H. Schwartz. New York, Amsterdam & Oxford: Elsevier, pp. 287–300.

Olds, J. (1976). Reward and drive neurons: 1975. In: *Brain-Stimulation Reward,* ed. A. Wauquier & E. T. Rolls. Oxford & New York: North-Holland/American Elsevier, pp. 1–27.

Plutchik, R. (1980). *Emotion: A Psychoevolutionary Synthesis.* New York: Harper & Row.

Racker, H. (1957). The meanings and uses of countertransference. *Psychoanal. Quart.,* 26:303–357.

Reiser, M. F. (1984). *Mind, Brain, Body.* New York: Basic Books.

_____ (1985). Converging sectors of psychoanalysis and neurobiology: Mutual challenges and opportunity. *J. Amer. Psychoanal. Assn.,* 33:11–34.

Rescorla, R. A. (1988). Pavlovian conditioning. It's not what you think it is. *Amer. Psychologist,* 43:151–160.

Rolls, E. T. (1976). The neurophysiological basis of brain-stimulation reward. In:

Brain-Stimulation Reward, ed. A. Wauquier & E. T. Rolls. Oxford & New York: North-Holland/American Elsevier, pp. 65–87.

Rosenblatt, A. D. (1985). The role of affect in cognitive psychology and psychoanalysis. *Psychoanal. Psychol.,* 2:85–97.

———— & Thickstun, J. T. (1977a). Energy, information, and motivation: A revision of psychoanalytic theory. *J. Amer. Psychoanal. Assn.,* 25:537–558.

———— ———— (1977b). *Modern Psychoanalytic Concepts in a General Psychology. Psychol. Issues,* Monogr. 42/43. New York: Int. Univ. Press.

Ross, E. D. (1984). Right hemisphere's role in language, affective behavior and emotion. *Trends Neurosci.,* 7(9):342–346.

Sandler, J. & Joffe, W. G. (1968). Psychoanalytic psychology and learning theory. In: *The Role of Learning in Psychotherapy.* CIBA Symposium. Boston: Little-Brown, pp. 274–287.

———— ———— (1969). Towards a basic psychoanalytic model. *Internat. J. Psychoanal.,* 50:79–90.

Schwartz, A. (1987). Drives, affects, behavior—and learning: Approaches to a psychobiology of emotion and to an integration of psychoanalytic and neurobiologic thought. *J. Amer. Psychoanal. Assn.,* 35:467–506.

———— (1988). Reification revisited: Some neurobiologically filtered views of "psychic structure" and "conflict." *J. Amer. Psychoanal. Assn.,* 36(Suppl.):359–385.

———— (1990a). On narcissism: An(other) introduction. In: *Pleasure Beyond the Pleasure Principle. The Role of Affect in Motivation, Development, and Adaptation, Vol. I,* ed. R. A. Glick & S. Bone. New Haven & London: Yale Univ. Press, pp. 111–137.

———— (1990b). To soothe or not to soothe—or when and how: Neurobiological and learning-psychological considerations of some complex clinical questions. *Psychoanal. Inq.,* 10:554–566.

———— (in prep.). Not art but science: Applications of neurobiology, experimental psychology, and ethology to psychoanalytic technique. II: Tactlessness and tact; safety, efficacy, and the "surface"; and the understanding of countertransference. *Psychoanal. Inq.* Material therein presented as a precirculated paper at the Fall Meeting of the American Psychoanalytic Association, New York, December 17, 1988.

Segal, H. (1964). *Introduction to the Work of Melanie Klein.* New York: Basic Books.

Snyder, S. H. (1980). Brain peptides as neurotransmitters. *Science,* 209:976–983.

Stechler, G. & Halton, A. (1987). The emergence of assertion and aggression during infancy: A psychoanalytic systems approach. *J. Amer. Psychoanal. Assn.,* 35:821–838.

Stellar, J. R. & Stellar, E. (1985). *The Neurobiology of Motivation and Reward.* New York, Berlin, Heidelberg & Tokyo: Springer-Verlag.

Strachey, J. (1934). The nature of the therapeutic action of psycho-analysis. *Int. J. Psychoanal.,* 15:127–159.

Tomkins, S. S. (1970). Affect as the primary motivational system. In: *Feelings and Emotions: The Loyola Symposium,* ed. M. B. Arnold. New York: Academic Press, pp. 101–110.

Wise, R. A. & Bozarth, M. A. (1985). Actions of abused drugs on reward systems in the brain. In: *Neurotoxicology,* ed. K. Blum & L. Manzo. New York & Basal: Marcel Dekker, pp. 111–133.

Barton Hall
St. Elizabeths Hospital
Washington, D.C. 20032

Experimental Studies of Higher Cortical Functions That Proceed Without Conscious Awareness

BRUCE E. WEXLER, M.D.

FROM A NEUROSCIENCE PERSPECTIVE, it is usually assumed that many higher cortical functions proceed without conscious awareness. In a recent overview paper in *Science,* Kihlstrom (1987) presented data from multiple sources that demonstrate that this is true for human beings as well as for other animals. From this perspective, consciousness, rather than unconscious mental life, is more the exception and the mystery. It is only from the perspective of the conscious subjective self, and a science based on that self in ways and to an extent that it is often unaware, that unconscious processes have been deemed unusual and mysterious.

The experimental study of unconscious cortical processes, however, has proved difficult and controversial despite their apparent ubiquity. The difficulty has centered on the problem of demonstrating that experimental stimuli are indeed processed without conscious awareness. Several laboratory methods have been created for this purpose, subjected to close critical scrutiny (cf. Eriksen, 1960; Holender, 1986) and then modified in response to the criticisms. This process, extending across nearly 40 years, has led to the development of methods that at least in recent years have enabled laboratory scientists to study unconscious cortical processes with confidence and rigor.

Dr. Wexler is Associate Professor of Psychiatry, Yale University School of Medicine.

The first major section of this paper will describe these experimental methods. Some effort will be made to provide a historical developmental context, with emphasis on the criticisms raised and the methodological responses to the criticisms. A complete history of the development of these methods, however, will not be attempted, nor will a comprehensive review of all studies with any of the individual methods. (Several such reviews of the earlier studies are available: Eriksen, 1960; Bevan, 1964; Dixon, 1971; Wolitzky & Wachtel, 1973; Erdelyi, 1974.) My goal, instead, will be to describe studies that are representative of different methodological approaches. These methods have made it possible to collect data on a number of interesting questions including: (1) What factors modulate the behavioral effects of information processed without conscious awareness? (2) How do conscious and unconscious information processing differ? (3) How do the brain processes on which conscious and unconscious processing depend differ from one another? My second main section will present existing data pertaining to these questions. The reader should appreciate in advance, however, that much of the data was collected in the process of developing and validating experimental methods rather than in studies designed specifically to explore one or another of the questions. Consequently they do not reflect systematic efforts to address the individual questions, and many gaps in knowledge will be evident. The concluding section will touch briefly on some more general issues.

Experimental Methods

Brief Exposure of Visual Stimuli

With this method test stimuli are presented for such a short time that subjects are not able to consciously identify them. In early studies the exposure time required for conscious identification was itself the dependent measure, and efforts were made to determine how this was affected by the nature of the stimuli or the subject's personality (e.g., McGinnes, 1949; Carpenter et al., 1956). This approach was abandoned, however, because of difficulty in determining whether subjects with higher thresholds were truly unconscious of the stimuli at stimulus durations that led to conscious identification by other subjects, or just required more certainty to offer a response. Similarly differences between subjects for emotion-laden and neutral stimuli might be due to different degrees of comfort in reporting the emotion-laden words. Subsequent studies therefore attempted to present stimuli to all subjects at such short exposure durations that they could not be identified, and then look for effects of these stimuli in subsequent behavior or in physiological responses. For example, Lazarus and McCleary (1951) conditioned a skin conductance response (GSR) to nonsense syllables and then presented the stimuli at such short durations that subjects could not correctly report whether they had seen a nonsense syllable that had been conditioned or one that had not. Skin conductance responses to conditioned stimuli were evident, however, even on those trials that subjects verbally reported having seen a nonconditioned stimulus. Taylor (1953) made similar observations using geometric forms rather than nonsense syllables as stimuli. In reviewing these studies Eriksen (1960) pointed out that subjects' physiological responses might have resulted from conscious partial recognition of the briefly presented test stimuli. While these partial cues did not provide enough information for correct verbal identification, and while not reflected in the categorical verbal responses involved in identifying the stimuli as one or another of a limited set of stimuli, they did provide enough conscious information to trigger the GSR. Eriksen (1960) suggested that in order to eliminate the possibility of responses based on partial cues from unidentifiable stimuli it is necessary to demonstrate that subjects can not reliably determine whether any stimulus at all has been presented, or reliably discriminate between experimental and control stimuli in a simple same-different discrimination task. These have become the methodological standards of more recent studies. For example, Spence and Holland (1962) presented the word "cheese" to subjects for 6.7 msec (1/150 of a second), then read the subjects a list of words and required after a short delay that they write down as many of the words as they could recall. The words to be recalled included some related to cheese by meaning (e.g., bread). At this short stimulus duration no subject could correctly identify the stimulus, and 80 percent of subjects could not reliably distinguish a blank slide from the "cheese" slide. When recall data from these 80 percent were evaluated, words associated with cheese by meaning were significantly more likely to be recalled than were other words. A control group who briefly closed their eyes prior to hearing the list of words were as likely to recall words unrelated to cheese as they were words related to cheese.

A study by Silverman et al. (1978) provides another example. In this study the message "Beating dad is wrong" or the message "Beating dad is ok" was presented for 4 msec to male college students immediately before their accuracy at dart-throwing was assessed. As the investigators predicted on the basis of psychoanalytic theory, performance was impaired following the first message and enhanced following the second. At the conclusion of the experiment, subjects were shown one of the messages at 4 msec exposures with four repetitions, then shown a second message in the same manner and asked to indicate whether the second message was the same as the first. Half the time the second message was the same and half the time it was different. With 20 such discriminations required, three subjects were correct 13 times, two subjects 12 times, one subject 11 times, eight subjects 10 times, three subjects 9 times, two subjects 8 times and three subjects 7 times. As a group they made 221 correct discriminations and 219 incorrect determinations.

In another study using this same paradigm, Ariam and Siller (1982) presented different groups of Israeli high school students with different messages four times each week over a six-week period. The message was presented twice, for 4 msec each time, in each of the four weekly sessions. The sessions were organized around mathematics classes, and subjects were told that the purpose of the study was to see if the messages would improve performance in mathematics. Two of the groups received different

Hebrew translations of the English phrase "Mommy and I are one." A third group received the Hebrew equivalent of "my teacher and I are one" and the fourth received the Hebrew translation of "people are walking in the street." The first two groups did significantly better on mathematics examinations at the end of the six-week period than did the second two groups. In order to rule out the possibility that subjects might be consciously aware of the stimuli another message (in Hebrew) "Sport is good for body and soul" was presented to a group of 28 students, with exposure duration beginning at 4 msec and increasing in 1 msec increments. Subjects were asked to report anything they saw with each presentation. The first report by any student of seeing anything other than a light flicker was the report of seeing a line at 12-msec exposure. The group mean first report of seeing anything was at 15.4 msec. The first report of seeing a letter was at 14 msec with the group mean at 25.7 msec. The first report of seeing the entire message was at 22 msec with the group mean at 33.8 msec. In an additional check at the conclusion of the actual study, subjects were told what the four experimental messages were and asked to guess which one they received. Eighteen of 65 students guessed correctly, which is not significantly greater than the 16.25 expected on the basis of a one-in-four guess rate.

Brief Exposure of Visual Stimuli Followed
by a Visual Mask

In this method stimuli are presented visually for a brief period and then followed by a second stimulus of longer duration. The presence of the second or masking stimulus limits conscious processing of the first stimulus. The masking stimulus is usually a pattern of visual noise, but it can be a meaningful stimulus. The presence of the masking stimulus makes it possible to present the initial stimulus for longer periods (10–40 msec) without leading to conscious awareness than is possible in the previously described visual studies, which did not employ a mask. This has the potential advantage of allowing more extensive processing of the initial stimulus, though this has not been established.

With this technique the ability of a subject to consciously identify a stimulus is influenced both by the duration of the stimulus and the time between presentation of the stimulus and presentation of the mask. In most studies these parameters are titrated individually for each subject to ensure that subjects can not reliably distinguish the presence or absence of an initial stimulus.

Holender (1986) has argued that because of differences in illumination between the threshold determinations (which usually precede the experiment itself) and the experiment itself, one can not be confident that subjects are not consciously aware of the initial stimuli during the experiment. A study by Humphreys et al. (1982), however, would seem to provide definitive reassurance on this point. Subjects were presented with a pattern mask initially, then two words in succession, and finally another pattern mask. The first word was always in lower case letters and the second in upper case. The two words could be the same (e.g., hair-HAIR), could be homophones (e.g., Hare-HAIR), could be graphemically related (e.g., harn-HAIR) or could be unrelated (e.g., food-HAIR). Exposure durations were set individually, based on initial studies that yielded approximately 40 percent identification of the second word when the first was unrelated. All four stimuli in each set were presented for the same duration and without interstimulus intervals. Exposure durations varied between 30 and 55 msec for different subjects. The subjects' task during the experiment proper was simply to report whatever words they saw. The initial word in each stimulus set was reported on only 1 percent of the trials, apparently being kept out of conscious awareness by the subsequent word and mask. Accuracy in identifying the second word in each stimulus set was, however, influenced by the nature of the first word. The second word was more accurately identified when the first word was the same or was a homophone than it was when it was unrelated or only graphemically related. This study addresses Holender's concern, since the inability to identify the first word was documented on essentially every trial during the experiment itself.

Marcel (1983) conducted a second study that would also seem to fully address the problem raised by Holender (1986). In this study four types of stimulus sequences were presented. In the first, a word was presented for 10 msec followed first by a pattern mask, and then 500 or 1000 msec later by a second word that was semantically related to the first (word1-pattern mask-500/1000 msec pause-word2). The second type of sequence was the same as the first except that the two words were unrelated. The third type was the same as the second except that last stimulus was a string of letters but not a word. The fourth type of stimulus sequence was the same as the others except that a blank field was presented for 10 msec prior to the pattern masking instead of the first word (blank-mask-pause-word or nonword). Prior to the presentation of each stimulus sequence, subjects were instructed to perform one of two tasks. On some trials they were asked to indicate whether the final stimulus was a word or not (lexical-decision task). On the other trials they were asked to indicate whether or not a word had been present prior to the mask (stimulus-detection task). Subjects proved unable to perform the detection task with accuracy significantly better than chance. However, the lexical-decision task was performed significantly more quickly if the initial word was semantically related to the final word than if it was not, despite the fact that subjects were not consciously aware if there was or was not an initial stimulus. Moreover, if the initial stimulus and pattern mask were presented repeatedly before presentation of the final word or nonword (up to 20 repetitions), the reaction time advantage on those trials when the initial and final words were semantically related increased steadily while performance on the decision task remained at chance levels.

A third study (Greenwald et al., 1989), conducted after the publication of Holender's article and designed specifically to address the methodological question raised in that review, yielded results parallel to those of Marcel (1983). Finally, a still more recent study by Niedenthal (1990) extended the earlier study by demonstrating processing of nonverbal emotion-related stimuli that were rendered unavailable to consciousness both by short (2 msec) stimulus exposure and backward masking. Subjects were asked to memorize and then recall a series of cartoon drawings presented via slide projector for two seconds. During the learning trials a picture of an actor's face was presented for 2 msec immediately prior to presentation of the cartoon (the cartoon itself served as the mask). Some faces expressed joy while others expressed disgust. When the cartoons were presented again for the recall task, they were again preceded by 2 msec exposures of the actors' faces. On some recall trials the faces preceding the cartoons expressed the same emotion as did the faces that had preceded the cartoons in the learning trials (e.g., joy-joy or disgust-disgust). On the other recall trials the faces preceding the cartoons expressed a different emotion from the face that had preceded the cartoons in the learning trials (e.g., joy-disgust or disgust-joy). In all cases the actors pictured during the recall task were different actors from those pictured during the learning trials. Thus in those recall trials on which the emotion preceding the cartoon was the same as had been the emotion preceding the cartoon during the learning trials, it was the nature of the emotion that was the same and not the actual face that was the same. Recall of the cartoons proved faster when the emotion expressed on the slides presented immediately prior to the cartoons was the same during training and recall than when the emotions differed. This was so despite the fact that on a simple detection task subjects were unable to determine whether a face or a blank slide was presented prior to presentation of the cartoon.

Presentation of Visual Stimuli at Low Levels of Intensity

In this method subjects are asked to watch as a line projected onto the center of an illuminated viewing screen is gradually dimmed, and to indicate when they are certain they can no longer see it. As described by Dixon (1958) this procedure is repeated four times, and the lowest illumination level at which subjects first report total disappearance of the line is taken as their individual threshold. This level is then further reduced for the experiment proper. Dixon presented emotion-evoking (e.g., penis, vagina) and neutral words (e.g., line, seven) at this subthreshold illumination level; he measured GSR and asked subjects to report whatever words came to mind and then to indicate whether they had seen anything. While no subject reported awareness of any stimuli, the GSR response was greater to the emotion-evoking than to the neutral stimuli. This experimental method has been used less than the others, perhaps because there is a smaller margin of error between subthreshold and superthreshold stimulation. In addition, Eriksen (1960) has raised questions about the availability of partial cues in this paradigm which have not been addressed in subsequent studies.

Presentation of Two Different Stimuli Simultaneously, One to Each Ear

In this method two different stimuli are presented simultaneously, one to each ear (dichotic listening). In early studies that attempted to use this method to study unconscious mental processes, investigators presented previously aversively conditioned nonsense syllables mixed among nonconditioned syllables in a string of stimuli to one ear, while a string consisting entirely of nonconditioned syllables was simultaneously presented to the other ear (Corteen & Wood, 1972; Dawson & Schell, 1982). Subjects were instructed to repeat aloud each syllable presented to the ear receiving exclusively nonconditioned stimuli, and to ignore the distracting syllables simultaneously presented to their other ear. Even when attention was thus directed away from the conditioned stimuli, these stimuli were associated with clear GSR responses. However, as Holender (1986) points out, this methodological approach is flawed because subjects are able to consciously identify stimuli presented to both ears in this type of dichotic stimulation, and consequently the conditioned stimuli are processed without conscious awareness only to the extent that subjects can limit normal processing capacity. More recent studies have addressed this problem by using the fused dichotic stimulation format developed by Halwes (1969) and Repp (1976). In this format two stimuli that differ in only one consonant are presented simultaneously, one to each ear (e.g., kill-till). Natural speech recordings of these utterances are digitized, and the distinguishing consonant from one utterance is cross-spliced onto a copy of the nondistinctive portion of the other utterance. Thus the members of each pair are identical save for the initial consonant. When the onsets of the two words or syllables in each stimulus pair are precisely aligned, the degree of auditory spectral and temporal overlap between them is so great that they fuse into a single auditory percept. Subjects report hearing only one word or syllable from each stimulus pair. They are unable to distinguish trials on which they receive fused dichotic pairs from those on which they receive the same stimulus in each ear (Repp, 1976). Moreover, even when told they would be receiving different words simultaneously in each ear, they were unable to selectively report words from only one ear (Wexler, 1988).

Two studies have recently been completed using fused dichotic pairs to study unconscious cortical processes. Both studies used special dichotic pairs that consisted either of a negative emotion-evoking word and a neutral word (e.g., kill-till) or a positive emotion-evoking word and a neutral word (e.g., fun-ton). Wexler et al. (1989) asked subjects to let their thoughts wander in response to the word they heard from each pair, during which time EEG activity in the alpha frequency over the right and left hemisphere frontal regions and EMG activity from the corrugator muscles that furrow the brow in a frown were recorded. After ten seconds of free association and physiological recording, subjects were asked to indicate what word they heard and to describe the thoughts they had in response to the word they heard. Even when the neutral word from the stimulus pair was heard, and the emotion-evoking word remained out of conscious awareness, emotion-specific changes were evident in both the EEG and the EMG recordings.

In the second recent study with the fused dichotic format, Bonano and Wexler (in press), using the same negative-neutral and positive-neutral pairs, required subjects to indicate what word they heard immediately after presentation of each pair. After presentation of 33 such pairs, subjects were given lists of five words corresponding to each stimulus pair. The lists contained both members of the stimulus pair and three foils differing from the stimulus words only in the first consonant (e.g., kill-till-gill-pill-dill). Subjects were required to rank the words according to the likelihood that they had heard the words during the period of stimulus presentation. As expected, words that subjects reported hearing immediately after stimulus presentation were usually ranked first. Of particular interest, however, was the fact that the members of the dichotic stimulus pairs that subjects were unaware of hearing were ranked significantly higher than were the words that had not been presented at all. This provides evidence of memory of stimuli processed without conscious awareness. A subsequent study with 80 additional subjects confirmed this finding (Bonano & Wexler, in press). Control studies ruled out the possibility that the words not heard were ranked higher than chance for reasons unrelated to their having been presented in the dichotic pairs, and provided additional evidence that subjects were not consciously aware of both words in the dichotic pairs.

Low Intensity Electrical Stimulation of the Skin

In this method, described by Libet et al. (1967), constant electrical pulses of varying amplitude are applied to the skin. Subjects are alerted that a stimulus may be given and know at what location it would be applied. Their task is to indicate: (1) they have felt a sensation even if very weak; (2) they have felt nothing at all; or (3) they are uncertain. The range of uncertain responses proved very small and false positives very rare. The subjects in this study were patients undergoing neurosurgical procedures who volunteered to participate in the study while in an unanesthetized state. Libet and colleagues were able to record evoked potentials directly from somatosensory cortex at precisely the location that, when directly stimulated, produced sensations attributed to the location on the skin where the experimental stimuli were applied. With this unusually high degree of experimental sensitivity and specificity, Libet et al. found evoked potentials on somatosensory cortex to cutaneous stimulation even when subjects were unaware of having been stimulated. Simultaneous recordings from scalp electrodes failed to reveal evoked potentials to the same stimuli, indicating the value of this unusual experimental opportunity. It must be noted, however, that even though the evoked potentials were recorded directly from the cortical surface it does not mean that they reflect activity in the cortex. They could represent volume conductance of electrical activity arising from the brain stem, and therefore not constitute evidence of higher brain function without conscious awareness. It is also true that the general applicability of this method may be limited first because the functional significance of the information processing reflected in the evoked potential is unknown, and second because it may require direct cortical recordings to detect evoked potentials to somatosensory stimuli that are below the threshold of conscious recognition.

Questions of Interest

What Factors Modulate Response to Information Processed Without Conscious Awareness?

As just reviewed, Libet et al. (1967) found that electrical stimuli applied to the skin that were too weak to cause a conscious perception still produced an evoked potential on the surface of the somatosensory cortex. When these same stimuli were repeated frequently enough, a conscious experience resulted. Apparently then, information processed without conscious awareness can be stored and combined with subsequent information that is also processed without conscious awareness. Marcel's study (1983) provides another example of this as the effect of unconsciously processed visually masked words on a subsequent lexical decision task was enhanced by multiple repetitions of the unconsciously processed stimulus prior to the task. Together these data suggest that summation of information from unconsciously processed stimuli can lead to new or enhanced behavioral effects.

Characteristics of the individual receiving the stimuli also seem able to alter the effects of unconsciously processed information. The effect of the word "cheese", when briefly presented below the exposure threshold necessary for conscious awareness, had a greater effect on subsequent recall of words associated by meaning with cheese in subjects who had not eaten for several hours than it did in subjects who had eaten recently (Spence & Ehrenberg, 1964). Spence and Ehrenberg interpreted this as showing that alterations in drive state can change the response to information processed without conscious awareness. Several other studies suggest that more enduring subject characteristics, as reflected in different personality organizations, can also influence responses to unconsciously processed information. In one of these studies Bonano and Wexler (in press) found that subjects defined by personality scales as repressors had less memory for unconsciously processed words than did nonrepressors. Shevrin et al. (1969) earlier reported similar findings with a different experimental procedure. Repressors, defined in their study on the basis of Rorschach responses, produced fewer verbal associations and showed lower evoked potential amplitudes in response to pictures presented visually at exposure durations that precluded conscious awareness, than did nonrepressors. Other studies have shown that repressors, again defined by personality scales, show greater physiological response but lower subjective response to consciously processed laboratory stimuli than do nonrepressors (e.g., Weinberger et al., 1979; Asendorf & Scherer, 1983; Cook, 1985). Wexler et al. (1989) recently extended these studies into the domain of unconscious processing. They found that repressors

had greater corrugator EMG responses to unconsciously processed negative emotion-evoking stimuli than they did to consciously processed negative stimuli, while the reverse was true for
nonrepressors. These studies did not examine evoked potentials,
and rather than contradicting the Shevrin et al. finding (1969),
suggest the possibility of enhanced response to unconsciously
processed stimuli in some physiological parameters, and diminished response in others as a function of personality organization.

The response to unconsciously processed information also
appears to depend on whether the experimental task calls for a
cognitive or an affective response. Kunst-Wilson and Zajonc
(1980) first demonstrated this by comparing affective and cognitive responses to a series of 11 irregular polygons each of which
was presented five times at 1 msec stimulus duration. In a
subsequent recognition test (cognitive task) subjects were only
right 48 percent of the time when asked to pick the polygons they
had seen from pairs of polygons consisting of one they had seen
and one they had not. However, when asked to indicate which of
each pair they liked better, subjects significantly more often
picked the one they had seen. Three subsequent studies (Seaman et
al., 1983a, 1983b, 1984) replicated this interesting finding. The
first of these studies found that affective discrimination remained
significant, and the cognitive recognition rate nonsignificant both
one day and one week after stimulus exposures, thereby providing
additional evidence for memory of unconsciously processed information. The last of these studies found that the affective discrimination rate remained constant at stimulus exposures of 2, 8, 12,
24, and 48 msec. Recognition accuracy, in contrast, remained at
chance levels with 2 and 8 msec exposures, but became more
accurate than affective discrimination at longer exposures. This
provides further demonstration of the independence of the affective and cognitive processing. Given this independence it is not
clear that the affective response to the stimulus itself was unconscious, nor is it clear just how one would demonstrate that it was
unconscious in the absence of the capacity for cognitive discrimination. It seems most likely, though, that there was no affective

response to the polygons when initially presented for 1 msec, but instead there was a positive affective response to the polygon that had previously been presented when it was presented for extended viewing along with a not previously seen polygon during the discrimination task. In other words, the positive affective response was not associated with the brief perception of the stimuli but rather with some sense of familiarity during the subsequent discrimination task. These data then suggest that in some circumstances unconsciously processed information may alter consciously experienced affective life without becoming available to conscious cognitive life.

How Do Conscious And Unconscious Information Processing Differ?

From a neuroscience perspective, the fact of information processing becoming conscious implies that a new set of physiological processes have become engaged. Engagement of a new set of physiological processes suggest, in turn, that there must be changes in the nature of information processing when that processing becomes conscious. Given the major role of consciousness in human experience and social exchange, it seems likely that these differences would be of at least moderate significance. The nature and extent of these differences is, however, an issue for empirical determination rather than theoretical inference or deduction. Laboratory studies conducted thus far point both to differences in the quality of conscious and unconscious information processing, and differences in the effects of conscious and unconscious information processing on subsequent behavior.

Bonano and Wexler (in press) found two interesting differences between conscious and unconscious information processing. As described above, they presented subjects with fused dichotic stimulus pairs consisting of an emotionally neutral word and either a positive emotion-evoking or a negative emotion-evoking word. Subjects were required to indicate what word they heard at the time of stimulus presentation, and then later to rank order a list containing the members of the dichotic pair and three highly similar foils with regard to the likelihood that they had heard the words during the period of stimulus presentation. Of words that subjects reported hearing at the time of stimulus presentation, and therefore were processed consciously, positive-emotion-evoking words were ranked as more likely to have been heard than were negative-emotion-evoking words. In contrast, of those words from each dichotic pair that subjects were not consciously aware of hearing, negative words were ranked as more likely to have been heard than were positive words. Apparently then, with conscious information processing there was enhanced memory of positive as opposed to negative-emotion-evoking words, while with unconscious processing the reverse was true.

In another aspect of this same study, Bonano and Wexler examined the effects of structured mental activity immediately after presentation of each stimulus pair on the subsequent recognition rankings. One group of subjects was instructed to sit quietly and let their thoughts wander for eight seconds after each dichotic pair. Another group was required to do the digit-symbol-substitution task during this time. A warning tone alerted both groups that the eight-second period was over and that a new stimulus pair would be presented in two more seconds. The "silence" condition and the digit-symbol condition proved to have different effects on conscious and unconscious processing. Moreover, the nature of these differences varied as a function of personality organization.

Requiring subjects to do the digit-symbol-substitution task led to lower recognition ranking of the words that subjects were aware of hearing at the time of stimulus presentation, as compared to the silence condition. This decreased memory of consciously processed words associated with the digit-symbol task was true for all personality types and for positive, negative, and neutral words alike. With regard to recognition rankings of the words that subjects were not consciously aware of hearing, the situation was quite different. Individuals defined by personality scales as high-anxious subjects had lower memory for unconsciously processed positive and negative words in the digit-symbol condition than they did in the silence condition. Memory of unconsciously processed neutral words, in contrast, did not differ in the two experimental conditions. While the digit-symbol task decreased memory of consciously processed words in the high-anxious regardless of the emotional salience of the word, with unconscious processing in these same subjects the effect of the digit-symbol task depended on the emotional salience of the word. Differences between conscious and unconscious processing were more striking in subjects defined as repressors. In these subjects, as in the others, the digit-symbol task led to decreased memory of consciously processed words whether the words were positive, negative, or neutral. In contrast, the digit-symbol task actually led to increased memory for unconsciously processed positive and negative words in repressors. A second study (Bonano & Wexler, in press) with 80 more subjects replicated these personality related differences between conscious and unconscious processing. Bonano and Wexler interpreted these observations as indicating that repressors employ active, unconscious, and internally focused mental processes to limit unconscious processing of emotion-evoking stimuli. The digit-symbol task interfered with these processes, leading to increased memory of the unconsciously processed stimuli. High-anxious subjects in contrast, they suggest, rely on externally focused mental activity to limit unconscious processing of emotion-evoking stimuli. Therefore, the digit-symbol task led to decreased memory in these subjects as compared to the silence condition.

Differences in the behavioral effects of conscious as compared to unconscious information processing, have been reported in studies conducted by Spence and Holland and by Silverman and his associates. Spence and Holland (1962), as described above, found that visual presentation of the word "cheese" at exposure durations too short to allow conscious registration facilitated recall of words associated to cheese by meaning. When the word cheese was presented for two seconds, an exposure duration at which the stimulus was consciously processed with ease, it had no facilitative effect on recall of cheese associates. Although the follow-up study in relatively food-deprived subjects (Spence & Ehrenberg, 1964) throws some doubt on this finding, since the facilitative effect on recall of cheese associates was found with both unconscious and conscious processing of the word cheese, another study by Spence (1964) provides additional support for the initial observation. In this study subjects were asked to recall a list of words that included the word cheese as well as words associated and unassociated to it by meaning. When the word cheese was forgotten or after it was recalled, equal numbers of cheese associates and nonassociates were recalled. However, prior to recall of the word cheese more cheese associates than nonassociates were recalled. This suggests that the word cheese activated memory traces of words associated to it by meaning more when it was unconscious than after it became conscious.

The work of Silverman and his associates with clinical populations has yielded clearer and more consistent evidence of the differential effects of conscious and unconscious information processing. In the first of these studies Silverman and Spiro (1968) presented to schizophrenic patients pictures of either an aggressive animal like a tiger, or a peaceful animal like a rabbit, and then looked for evidence of thought disorder in subsequent word association and recall tasks. When the picture of the tiger was presented for only 4 msec it led to an increase in thought disorder, while presentation of a blank slide or of a peaceful animal did not. When the picture of the tiger was presented for 10 seconds it had no such effect. Subsequent studies (Silverman & Candell, 1970; Rutstein & Goldberger, 1973) also found that supraliminal presentations failed to effect subsequent behavior while subliminal presentations of the same stimuli did. Apparently then, engagement of the neurophysiological processes associated with conscious experience can limit the behavioral effects of some types of information processing.

How Do The Brain Processes On Which Conscious and Unconscious Processing Depend Differ From One Another?

Wexler and colleagues (1989) have explored neuroanatomic differences between conscious and unconscious information processing with two different experimental approaches. The first uses lateralized physiological responses to compare relative right and left cerebral hemisphere activation with conscious and unconscious information processing. The second compares physiological responses to consciously and unconsciously processed stimuli as a function of which cerebral hemisphere initially processes the stimuli.

Lateralized cerebral activation was measured by recording EEG activity in the alpha frequency band from scalp leads over left and right frontal regions, and by recording EMG activity from the left and right sides of the face (corrugator muscle regions). Both physiological measures indicated that when emotion-evoking stimuli were presented as part of a fused dichotic stimulus pair and processed without conscious awareness, there was greater right than left hemisphere activation. When the same stimuli were processed consciously the opposite pattern prevailed and there was greater left than right hemisphere activation (Wexler et al., 1989). These data point to a special role for left hemisphere systems in conscious awareness.

Further support for this view emerged when physiological responses to consciously and unconsciously processed emotion-evoking stimuli were evaluated as a function of the hemisphere that initially processed the stimuli. With fused dichotic stimulus presentation, information from each ear is initially processed exclusively in the contralateral cerebral hemisphere. (Studies in a variety of different types of neurological patients indicate this is so despite the presence of minor ipsilateral auditory afferent pathways. Apparently with dichotic stimulation these minor ipsilateral pathways are rendered nonfunctional. See Wexler [1988] for a fuller discussion.) Wexler et al. (1989) found that when negative-emotion-evoking stimuli were presented to the right hemisphere both true low-anxious subjects and repressors showed equal corrugator EMG responses to consciously and unconsciously processed stimuli. When the same stimuli were presented to the left hemisphere, however, repressors showed a 14-fold greater EMG response when they processed the negative-emotion-evoking stimuli unconsciously than they did when they processed the stimuli consciously. True low-anxious subjects showed the opposite pattern, with a ten-fold greater response with conscious than with unconscious processing. The fact that this interaction between personality organization and conscious vs. unconscious processing only emerged when the emotion-evoking stimuli were presented to the left hemisphere again suggests a special role for left hemisphere neural processes in consciousness. The Bonano and Wexler (in press) study adds more support for this view. As described above, when repressors were required to do the digit-symbol-substitution test immediately after the presentation of each dichotic pair, subsequent recognition of the unconsciously processed emotion-evoking stimuli was actually enhanced. This was seen as consistent with the view of repression as an active unconscious process that requires mental resources. Doing the digit-symbol-substitution test apparently competed for those resources and thereby interfered with repression. Based on the observation that differences between repressors and true low-anxious subjects in corrugator response to unconsciously processed negative-emotion-evoking stimuli only appeared when those stimuli were presented to the left hemisphere, Wexler hypothesized that the digit-symbol-substitution task interfered with repression because it drew on left hemisphere processing resources. Bonano and Wexler (in press) evaluated this hypothesis in their replication study by replacing the digit-symbol-substitution task with either a counting-backward task or a geometric figures mental rotation task. The first was chosen because it draws primarily on specialized left hemisphere processing systems while the second was chosen since it draws primarily on specialized right hemisphere processing systems. When repressors were required to count backward (left hemisphere task) immediately after presentation of each dichotic pair, subsequent recognition of the unconsciously processed emotion-evoking stimuli was enhanced just as had been the case with the digit-symbol-substitution task. Requiring subjects to do the mental rotation task (right hemisphere task) had no such effect. The data from both the Wexler et al. (1989) and the Bonano and Wexler (in press) studies, then, converge to suggest a special role for the left cerebral hemisphere in regulating access to consciousness and in consciousness-mediated responses to emotion-evoking stimuli. It must be kept in mind, however, that both studies used verbal stimuli that required specialized left hemisphere processes. It is possible that with test stimuli requiring specialized right hemisphere processing, the right hemisphere plays a special role in consciousness-related responses.

Concluding Remarks

A variety of experimental techniques are now available to study information processing that proceeds without conscious awareness. These techniques will make it possible to follow leads based on information gleaned from clinical treatment situations, and learn more about the processes that regulate responses to information processed unconsciously. Studies conducted thus far suggest that a subject's receptive physiological state as well as more enduring personality traits are relevant. These techniques will also make it possible to learn more about the differences between brain processes associated with conscious and unconscious information processing. Studies thus far suggest a specialized role for left hemisphere processes in modulating consciousness-related behaviors, although this may be due to the fact that these studies used language-related stimuli. These initial studies used the fused dichotic listening format, but brief visual presentations in a single visual field also permit single hemisphere presentations, and future studies should use both techniques to allow convergent validation of findings. Studies like these will make more connections between our understandings of brain processes and mental processes in general. In particular such studies have the potential to help delineate the neural processes most closely associated with the mystery of conscious experience.

Future studies will also be able to further define the differences between conscious and unconscious information processing. From the perspective of our conscious subjective selves, this will enable us to learn more about the qualities of the silent partners with which we function in tandem. From a broader scientific perspective, such studies will enable us to learn more about the special qualities of conscious information processing. This information may help us understand the special role of consciousness in human development and function, as well as in the overall scheme of evolution. While the existing experimental data relevant to these questions are very limited, they do suggest some intriguing hypotheses. The studies by both Spence and Silverman and their associates suggest that information processing that proceeds without conscious awareness may have wider ranging effects on other mental processing than does conscious information processing. Consciousness would seem from this to limit and contain brain and mental operations. Conscious experience may itself be the subjective awareness of these boundaries. Such awareness may take the form of "I am thinking about this now." Such conscious experiences also provide the basis for interpersonal exchange of information and cooperative activities. Such social exchanges, and larger-group social processes, influence the conscious state of mind. These data and observations suggest the hypothesis, then, that consciousness may be in part a mechanism through which social interactions can limit the brain processes of individuals and thereby facilitate group enterprise.

REFERENCES

Ariam, S. & Siller, J. (1982). Effects of subliminal oneness stimuli in Hebrew on academic performance of Israeli high school students: Further evidence on the adaptation-enhancing effects of symbiotic fantasies in another culture using another language. *J. Abnormal Psychol.*, 91:343–349.

Asendorph, J. & Scherer, K. (1983). The discrepant repressor: Differentiation between low anxiety, high anxiety, and repression of anxiety by autonomic-facial-verbal patterns of behavior. *J. Personality & Soc. Psychol.*, 45:1334–1346.

Bevan, W. (1964). Subliminal stimulation: A pervasive problem for psychology. *Psychol. Bull.*, 61:81–99.

Bonano, G. & Wexler, B. (in press). Repressors, blunters and the processing of competing dichotic stimuli. *J. Personality.*

Carpenter, B., Weiner, M., & Carpenter, J. (1956). Predictability of perceptual defense behavior. *J. Abnormal & Soc. Psychol.*, 52:380–383.

Cook, J. (1985). Repression-sensitization and approach avoidance as predictors of response to a laboratory stressor. *J. Personality & Soc. Psychol.*, 49:759–773.

Corteen, R. & Wood, B. (1972). Autonomic responses to shock-associated words in an unattended channel. *J. Exp. Psychol.*, 94:308–313.

Dawson, M. & Schell, A. (1982). Electrodermal responses to attended and non-attended significant stimuli during dichotic listening. *J. Exp. Psychol.: Human Perception & Performance*, 8:315–324.

Dixon, N. (1958). The effect of subliminal stimulation upon autonomic and verbal behavior. *J. Abnormal & Soc. Psychol.*, 57:29–36.

——— (1971). *Subliminal perception: The nature of a controversy.* London: McGraw-Hill.

Erdelyi, M. (1974). A new look at the new look: Perceptual defence and vigilance. *Psychol. Rev.*, 81:1–25.

Eriksen, C. (1960). Discrimination and learning without awareness: A methodological survey and evaluation. *Psychol. Rev.*, 67:279–300.

Greenwald, A., Klinger, M., & Liu, T. (1989). Unconscious processing of dichoptically masked words. *Memory & Cognition*, 17:35–47.

Halwes, T. (1969). Effects dichotic fusion on the perception of speech. Status report on speech research. New Haven: Haskins Laboratories.

Holender, D. (1986). Semantic activation without conscious identification in dichotic listening, parafoveal vision, and visual masking: A survey and appraisal. *Behavioral & Brain Sciences*, 9:1–66.

Humphreys, G., Evett, L., & Taylor, D. (1982). Automatic phonological priming in visual word recognition. *Memory & Cognition*, 10:576–590.

Kihlstrom, J. (1987). The cognitive unconscious. *Science*, 237:1445–1452.

Kunst-Wilson, W. & Zajonc, R. (1980). Affective discrimination of stimuli that cannot be recognized. *Science*, 207:557–558.

Lazarus, R. & McCleary, R. (1951). Autonomic discrimination without awareness: A study of subception. *Psychol. Rev.*, 58:113–122.

Libet, B., Alberts, W., Wright, E., Jr, & Feinstein, B. (1967). Responses of human somatosensory cortex to stimuli below threshold for conscious sensation. *Science*, 158:1597–1600.

Marcel, A. (1983). Conscious and unconscious perception: Experiments on visual masking and word recognition. *Cogn. Psychol.*, 15:197–237.

McGinnes, E. (1949). Emotionality and perceptual defense. *Psychol. Rev.*, 56:244–251.

Niedenthal, P. (1990). Implicit perception of affective information. *J. Exper. Soc. Psychol.*, 26:505–527.

Repp, B. (1976). Identification of dichotic fusions. *J. Acoustical Society America*, 60:456–469.

Rutstein, E. & Goldberger, L. (1973). The effects of aggressive stimulation on suicidal patients: An experimental study of the psychoanalytic theory of suicide. *Psychoanal. & Contemp. Sci.*, 2:157–174.

Seamen, J., Brody, N., & Kauff, D. (1983a). Affective discrimination of stimuli that are not recognized: Effects of shadowing, masking, and cerebral laterality. *J. Exper. Psychol.: Learning, Memory, & Cognition,* 9:544–555.

———— ———— ———— (1983b). Affective discrimination of stimuli that are not recognized: II. Effect of delay between study and test. *Bull. Psychonomic Society,* 21:187–189.

———— ———— ———— (1984). Critical importance of exposure duration for effective discrimination of stimuli that are not recognized. *J. Exper. Psychol. Learning, Memory, & Cognition,* 10:465–469.

Shevrin, H., Smith, W. & Fritzler, D. (1969). Repressiveness as a factor in the subliminal activation of brain and verbal responses. *J. Nerv. Ment. Dis.,* 149:261–269.

Silverman, L., & Candell, P. (1970). On the relationship between aggressive activation, symbiotic merging, intactness of body boundaries and manifest pathology in schizophrenics. *J. Nerv. Ment. Dis.,* 150:387–399.

———— Ross, D., Adler, J. & Lustig, D. (1978). Simple research paradigm for demonstrating subliminal psychodynamic activation: Effects of oedipal stimuli on dart-throwing accuracy in college males. *J. Abnormal Psychol.,* 7:341–357.

———— & Spiro, R. (1968). The effects of subliminal, supraliminal and vocalized aggression on the ego functioning of schizophrenics. *J. Nerv. Ment. Dis.,* 146:50–61.

Spence, D. (1964). Conscious and preconscious influences on recall. Another example of the restricting effects of awareness. *J. Abnormal & Soc. Psychol.,* 68:92–99.

———— & Ehrenberg, B. (1964). Effects of oral deprivation on responses to subliminal and supraliminal verbal food stimuli. *J. Abnormal & Soc. Psychol.,* 69:10–18.

———— & Holland, B. (1962). The restricting effects of awareness: A paradox and an explanation. *J. Abnormal Psychol.,* 64:163–174.

Taylor, F. (1953). The discrimination of subliminal visual stimuli. *Canad. J. Psychol.,* 7:12–29.

Weinberger, D., Schwartz, G., & Davidson, R. (1979). Low anxious, high anxious, and repressive coping styles: Psychometric patterns and behavioral and psychological responses to stress. *J. Abnormal Psychol.,* 88:369–380.

Wexler, B. (1988). Dichotic presentation as a method for single hemisphere stimulation. In *Handbook of Dichotic Listening,* ed. H. Hugdahl. London: Wiley, pp. 85–115.

———— Warrenburg, S., Schwartz, G., & Jammer, L. (1989). EMG, EEG and heart rate responses to unconsciously processed emotion-evoking stimuli. Neuropsychologia. Presented at the Meeting of the American Psychiatric Association, San Fransisco.

Wolitzky, D. & Wachtel, P. (1973). Personality and perception. In *Handbook of General Psychology,* ed. B. Wolman. Englewood Cliffs, NJ: Prentice-Hall, pp. 826–857.

Department of Psychiatry
Yale University School of Medicine
34 Park St.
New Haven, CT 06519

Epilogue

A S GILBERT LONG AGO CONCLUDED, "Things are seldom what they seem," and one might in attempting a summary of this volume note, even underscore, an intriguing present-day irony. Despite its political, academic, and economic ascendancy, "biological psychiatry" has yet to furnish us with data describing decisively the etiology, pathophysiology, and eradication of a single diagnostic entity (cf. Gold et al., 1988a,b). On the other hand, converging lines of evidence and derivative models from the laboratories of some the world's leading bench neuroscientists hold the now very real promise of explaining the basic nature, cause, and cure of "psychic conflict" and the character disorders it engenders.

More explicitly, current ideas and data from neurobiology, experimental psychology, and ethology can help us achieve three different but closely related objectives—first, the reconceptualization and even the laboratory investigation of such clinical and theoretical keystones as motivational processes, consciousness, and "the unconscious," tasks which Hadley, Olds, and Wexler respectively undertake; second, the systematic and biologically sound reformulation of our understanding of symptom formation and therapeutic effect, ends to which Cooper contributes in his exploration of links between the fields of associative learning and psychoanalysis; and, finally, the placing of analytic technique on a solidly scientific foundation, a goal that Schwartz attempts to approach in his examination of affect, nonverbal communication, and determination of interpretive focus and timing.

Neuroscientifically based "reinterpretation" and "reexplanation" of long-familiar conceptual notions, clinical phenomena, and technical tasks may in turn help to ease a scientific—and, indeed, political—predicament that increasingly confronts psychoanalytically anchored treatments in these times of escalating pressure for accountability within the biomedical world. Although even the manufacturers of an over-the-counter pharmaceutical must establish that the therapy is rational, safe, and effective before the medication may appear on pharmacy shelves, analysis and derivative "talking cures" have not to date met these widely accepted and yet rather minimal standards. However, as the articles presented here suggest, we can today at least begin to argue cogently that the seemingly mystical data and processes of the consulting room mesh smoothly and elegantly with what bench investigators observe, and these first steps toward "rationality" will no doubt facilitate future empirical demonstration of "safety" and "efficacy": if we define our variables in biologic and physical terms, we then have an easier time measuring our results.

Obviously, a central element in the scientific refurbishment of analysis is a sort of linguistic translation—the finding within the vocabularies of neuroscience, ethology, and experimental psychology of concepts, models, and findings which closely fit clinical observations—and a future issue of this journal will continue these efforts with papers which examine, for instance, relationships between neural maturation and psychologic development; conceptualization of the role of analytic psychotherapy in management of medicated psychotics; and the neurobiologic understanding of tact, countertransference, and other aspects of interpretive technique.

In closing, and from a historical perspective, one might highlight still another intriguing and rather delicious irony. Two seemingly distant and often mutually hostile disciplines, behaviorism and psychoanalysis, early in their development shared one epistemologic tenet—the view that the brain was a "black box" whose neuroscientific understanding was at best irrelevant to each field's lines of investigation. Today, however, as the cellular neurobiology of learning advances in ways only dreamed of two decades ago, neuroscience, analysis, and experimental psychology appear, as Kandel (1979) suggested, to have a great deal to offer each other. The walls of the "black box" — and thus the underlying kinship of three ostensibly unrelated scientific approaches — become ever more transparent and clear.

REFERENCES

Gold, P. W., Goodwin, F. K., & Chrousos, G. P. (1988a). Clinical and biochemical manifestations of depression. Relation to the neurobiology of stress. I. *New Eng. J. Med.*, 319:348–353.
_____ _____ _____ (1988b). Clinical and biochemical manifestations of depression. Relation to the neurobiology of stress. II. *New Eng. J. Med.*, 319:413–420.
Kandel, E. R. (1979). Psychotherapy and the single synapse. *New Eng. J. Med.*, 301:1029–1037.

Andrew Schwartz, M.D
Issue Editor